Spreadsheet Chemistry

O. Jerry Parker
Gary L. Breneman

Department of Chemistry and Biochemistry
Eastern Washington University

Prentice Hall, Englewood Cliffs, New Jersey 07632

Library of Congress Cataloging-in Publication Data

Parker, O. J. (O. Jerry)
Spreadsheet chemistry / O. Jerry Parker, Gary L. Breneman.
p. cm.
Includes index.
ISBN 0-13-835562-2
1. Chemistry--Data processing. 2. Electronic spreadsheets.
I. Breneman, G. L. (Gary L.) II. Title
QD39.3.E46P34 1991
542'.8--dc20 90-45643
CIP

Editorial/production supervision: ***bookworks***
Manufacturing buyers: Paula Massenaro and Lori Bulwin

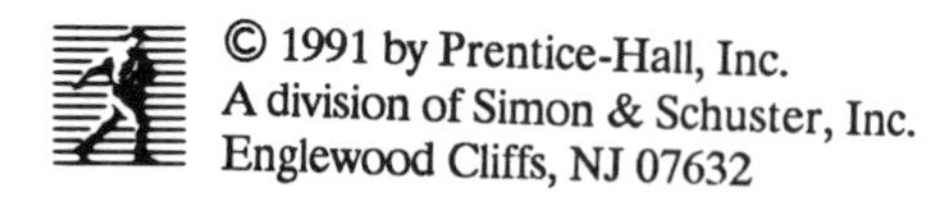

Printed in the United States of America

10 9 8 7 6 5 4 3 2 1

ISBN 0-13-835562-2

Prentice-Hall International (UK) Limited, *London*
Prentice-Hall of Australia Pty. Limited, *Sydney*
Prentice-Hall Canada Inc., *Toronto*
Prentice-Hall Hispanoamericana, S.A., *Mexico*
Prentice-Hall of India Private Limited, *New Delhi*
Prentice-Hall of Japan, Inc., *Tokyo*
Simon & Schuster Asia Pte. Ltd., *Singapore*
Editora Prentice-Hall do Brasil, Ltda., *Rio de Janeiro*

Contents

To the Student

All of the professions have been noticeably altered by the ubiquitous presence of the personal computer and contemporary software. Spreadsheet software represents a high level computer language that has proven to be very useful in the financial and business community. The attributes that make this software so valuable to these professional groups are exactly the features necessary for the mathematical interpretation of a wide variety of chemical and physical phenomena. The ability to explore options and to play "what-if" games by varying the theoretical assumptions, conditions, and chemical concentrations, coupled with the ease of obtaining the resulting calculations, has led to a computational revolution in the chemical sciences. Modern spreadsheet programs contain most of the mathematical functions needed by scientists. Spreadsheets represent an extremely fast and efficient method for inputting data for both numerical calculations and graphical presentations of the results. Recent software innovations have produced the intuitive graphical interface with pull-down menus, icons, and dialog boxes that make it easy, even for people with limited computer experience, to operate and produce results with spreadsheets. The skills needed to function effectively with spreadsheets are easily and quickly acquired by patterning your approach from examples of spreadsheets that accomplish familiar calculations. This is the approach that will be presented in this book. You should increase the level and variability of your computing repertoire by using the examples that are presented. Soon, you will be capable of extending the approaches used in the early chapters of this book, and with this newly acquired skill you will commence to design your own approaches and solutions to very complicated chemical systems.

Personal computers provide a means for quickly and efficiently calculating numerical results from theory, experimental data, and from questions posed about changes in the conditions that govern chemical systems. They allow an emphasis on the issues that really count to a chemist. Their use gives immediate access to results from theoretical models that formerly required hours of human computations. This speed and immediate access will encourage you to pose critical questions about the consequences of changes in the conditions that govern the behavior of chemical systems.

Spreadsheet Chemistry allows numerical exploration with the theory and principles that govern electron distributions in atoms and molecules; the thermodynamic quantities of enthalpy, entropy, and free energy; ideal and real gas behavior; systems at and kinetically approaching chemical equilibrium; titration curves for both acid-base and oxidation-reduction reactions; and the solubility of ionic solids. The computational speed and ease with which complex mathematical functions can be applied is utilized for calculating empirical and molecular formulas, the distribution of molecular velocities for gases, the vapor pressure of substances at various temperatures, atomic spectra and energy levels, and the nuclear binding energy of isotopes.

Many of the worksheets take advantage of the ease with which data can be graphed with the spreadsheet software. The graphical, visual presentation of data and calculations is one of the most powerful facets of spreadsheets as a problem-solving media. The ability to solve chemical problems is facilitated by the use of spreadsheets as a high level language that is more intuitive than other computer languages. The worksheets and charts developed in this text have been produced with Microsoft® Excel, but any spreadsheet software can be used in conjunction with *Spreadsheet Chemistry*.

Spreadsheet Chemistry is arranged for use with a standard general chemistry text such as *Chemistry: The Central Science* by Brown, LeMay, and Bursten. It is designed to enhance your problem-solving skills as well as your innate understanding of important chemical principles. Personal computers are an indispensable tool for the chemist. Ease of operation and endless applications to problems of chemical interest have made spreadsheets a necessity for utilizing the full potential of personal computers.

Introduction

With the current widespread access to personal computers comes a real opportunity to ease the mathematical burden associated with learning chemistry. The computer can continue the trend that calculators started a number of years ago by making routine that which in the past has been very long and tedious. The difficulty in learning much of chemistry has been in the mathematics, which can be time consuming and/or difficult to understand. This can obscure many chemical principles that are themselves simple, as well as powerful. The computer allows us to concentrate on these principles and puts mathematics in its place, as a tool instead of a hindrance, in the study of these chemical principles.

Spreadsheet programs were originally developed for personal computers for business use and, along with word processing programs, are responsible for the tremendous growth of computer use in our society. These spreadsheet programs include powerful mathematical functions that are also used in the scientific world. Thus, spreadsheets were discovered to be of great use in science. For the last several years, almost every issue of the *Journal of Chemical Education* has contained articles on using spreadsheets to do chemical calculations.

This book will introduce you to the power of the personal computer and spreadsheet programs while allowing you to look at chemistry in a light not seen by previous generations of chemistry students. You will be able to do numerous calculations very rapidly and, more importantly, experiment quickly with systems by choosing your own parameters and immediately seeing the results. This will let you consider the kind of "what if?" questions that spreadsheets are so famous for answering. Trends, hidden in the equations, become apparent, especially when put in graphical form, as seen in the many charts in this book.

The approach used by this book is to gradually introduce you to ever more powerful worksheets as you go through the various chapters covering many of the usual topics in general chemistry. Each chapter will give you a brief discussion of a topic (you should always refer to your regular chemistry text, such as Brown, LeMay, and Bursten referenced at the beginning of each chapter, for a more detailed discussion) and then provide you with worksheets that you can enter into your computer for solving a variety of problems. In many cases, charts will be made so you can see your results in graphical form. You will be given changes to make in these worksheets so they can handle additional problems. If you study the worksheets and the descriptions of how they work, you can also learn how to write your own simple worksheets, although this is not the main purpose of the book. The mathematics behind some worksheets may not be described in complete detail, such as the integration (a calculus concept) in Chapter 12 on Gas Kinetic-Molecular Theory. The chemistry itself is not at all obscured by this and in fact the results of the calculations make the idea of a velocity distribution much more meaningful.

The early chapters in the book describe how to enter the worksheets and charts into your computer and use them. Appendices A, C, and D summarize these procedures. Saving your worksheets and charts on disk is described in Appendix E. Important mathematical methods are considered in Appendices B, F, and G.

Numbering of *Worksheets, Lists, Tables, and Figures* all start with the chapter number and then each type is numbered sequentially through that chapter, with a period separating the two numbers, e.g., Worksheet 7.3 and Table 18.1. Occasionally, special types of worksheets will have a letter appended to the number, such as Worksheet 2.3F (a formula sheet), or Worksheet 13.12M (a function macro sheet).

Numbering of *Charts* is *not* sequential. Charts are numbered with the *same* numbers as the worksheets from which they were made (e.g., Chart 7.6 was made from Worksheet 7.6 and there may be no Charts 7.1 through 7.5). If more than one chart comes from the same worksheet, then the letters A, B. C., etc., are added to the numbers (e.g., Chart 9.4B is the second chart made from Worksheet 9.4).

Hardware and Software Requirements and Availability

The worksheets and charts described are for the IBM version of Microsoft Excel. IBM or IBM-compatible computers are the most common type of microcomputers currently available at colleges and universities. Excel is perhaps the easiest spreadsheet program to learn and use because of its graphical user interface. It allows the simultaneous display of worksheets and charts, so changes in one will immediately be shown in the other. It also allows this book to be used with the Macintosh version of Excel with no modifications. In fact, all the figures of worksheets and charts in this book were printed from the Macintosh version.

Excel requires a hard disk drive to run on an IBM type computer since it runs under Microsoft Windows. It can be run on an XT (8088) level computer (640K RAM) but will run faster on AT (80286) or higher level (80386, 80486) machines. The Macintosh version of Excel runs in a similar manner from a hard disk drive on any Macintosh computer with one Mb RAM. Worksheets and charts can be saved (Appendix E) on the hard drive or on a formatted floppy disk. A printer is not required but is a useful accessory for printing copies of your results.

The tearout card inside the back cover of this book allows you to order a promotional copy of the IBM version of Excel (includes a runtime version of Windows) that will run most of the sheets described in this book. Data disks containing all the sheets and charts in this book for both the IBM and Macintosh versions of Excel can also be ordered on this card

Acknowledgments

Several people have contributed in numerous ways to the preparation of this book. We would like to thank several colleagues who have reviewed the manuscript and help us by sharing their insights and collective experience.

Steven D. Brown	University of Delaware
G. A. Crosby	Washington State University
Dan Dill	Boston University
Peter Gold	Pennsylvania State University
Peter Lykos	Illinois Institute of Technology
R. Kenneth Marcus	Clemson University
David M. Whisnant	Woffard College
Richard York	Wittenberg University
David Zellmer	California State University-Fresno
J. W. Zubrick	Hudson Valley Community College

We heartily appreciate the assistance of many people at Prentice Hall but especially owe a debt of gratitude to Dan Joraanstad, our chemistry editor, who has furnished cheerful encouragement throughout the project. We wish to express our gratitude to Lisa Garboski who has competently served as the production supervisor and substantially contributed to the design of the text. A special debt of thanks from O. J. P. to Dana Parker for proofreading his portion of the manuscript.

O. Jerry Parker
Department of Chemistry and Biochemistry
Eastern Washington University, Cheney

Gary L. Breneman
Department of Chemistry and Biochemistry
Eastern Washington University, Cheney

1

Using Spreadsheets

Chemistry is the study of the material world that we inhabit. Chemists have developed intellectual explanations for our material environment that have resulted in a deeper understanding of the fundamental processes that are the primary concern of such diverse fields as chemical engineering, medicine, molecular biology, material science, and environmental science. While many of the significant developments of chemistry have involved practical applications, a broad definition of the discipline embodies a study of the materials of the universe and the changes that these materials undergo. Chemistry is an experimental science that involves characterizations, classifications, measurements, hypotheses, models, theories, laws, and predictions. Mathematical calculations, functions, and models are an integral aspect of the study of modern chemistry. Your understanding of chemistry will be enhanced by developing your intuitive insight into the consequences of varying parameters that govern the behavior of chemical systems. The techniques of organizing data, of data manipulation, interpretation, and the presentation of data in both graphical and tabular form will serve you well in other aspects of your professional career, as well as in your understanding of the physical world.

This book will introduce you to the power of the personal computer and the associated spreadsheet software that is now available for individual students, teachers, scientists, and engineers. This awesome computing power that is now available to the common individual is a remarkable feature of life during the information age. The techniques and applications presented in this book will allow you to accomplish calculations and display the results in graphical form on a monitor or as a hard copy on paper. The products of your work will have a very professional appearance in comparison with presentations produced with pen and paper. The type of computational activity that we will undertake was possible only on large, expensive computers some twenty years ago. If you choose to take advantage of these tools, your class work will be easier, and your opportunities for exploring the excitement of chemistry will be greatly enriched.

Spreadsheet software represents a high level computer language that has proven to be very useful in the financial and business community. The attributes that make this software so valuable to these professional groups are exactly the features necessary for the mathematical interpretation of a wide variety of chemical and physical phenomena. The ability to explore options and to play "what-if" games by varying the theoretical assumptions, conditions, and chemical concentrations, coupled with the ease of obtaining the resulting calculations, has led to a computational revolution in the chemical sciences. Modern spreadsheet programs contain most of the mathematical functions needed by scientists. Spreadsheets represent an extremely fast and efficient method for inputting data for both numerical calculations and graphical presentations of the results. Recent software innovations have produced the intuitive graphical interface with pull-down menus, icons, and dialog boxes that make it easy, even for people with limited computer experience, to operate and produce results with spreadsheets. The skills needed to function effectively with spreadsheets are easily and quickly acquired by patterning your approach from examples of spreadsheets that accomplish familiar calculations. This is the approach that will be presented in this book. You should increase the level and variability of your computing repertoire by using the examples that are presented. Soon, you will be capable of extending the approaches used in the early chapters of this book, and with this newly acquired skill you will commence to design your own approaches and solutions to very complicated chemical systems.

Numbers, calculations, functions, and models provide the framework upon which modern chemistry is based. Before we proceed with some of the more exciting examples of spreadsheet use, it will be necessary to invest some time in becoming familiar with the operating procedures that are frequently needed in order to effectively utilize spreadsheets. Familiar examples will be used in the first few chapters, with the major focus on operating procedures for spreadsheet applications. The first example will involve the conversion of degrees Celsius to degrees Fahrenheit, with the results presented in both tabular and graphical formats. (BLB Chap. 1)

Spreadsheet Calculations

With a new worksheet on the monitor of your computer and Microsoft® Excel or a similar spreadsheet program operating, you should produce an exact copy of Worksheet 1.1, which is our first example. The spreadsheet on the left represents what you see and the one on the right shows the formulas. Most of the entries in this worksheet may be made by using the Fill Down command under Edit on the menu bar. The four cells that you need to fill from the keyboard are highlighted by bold type. If you need directions for entering the information in the individual cells, you should refer to the specific directions that are given in Table 1.1. If you are unfamiliar with the general characteristics of spreadsheets, you should refer to Appendix A on Using Microsoft® Excel. Appendix E, Saving Files With Excel, has information on saving worksheets and charts when you are using Microsoft® Excel.

	A	B
1	Celsius ->	Fahrenheit
2	0	32
3	10	50
4	20	68
5	30	86
6	40	104
7	50	122
8	60	140
9	70	158
10	80	176
11	90	194
12	100	212

	A	B
1	**Celsius ->**	**Fahrenheit**
2	**0**	**=9/5*A2+32**
3	**=A2+10**	=9/5*A3+32
4	=A3+10	=9/5*A4+32
5	=A4+10	=9/5*A5+32
6	=A5+10	=9/5*A6+32
7	=A6+10	=9/5*A7+32
8	=A7+10	=9/5*A8+32
9	=A8+10	=9/5*A9+32
10	=A9+10	=9/5*A10+32
11	=A10+10	=9/5*A11+32
12	=A11+10	=9/5*A12+32

Worksheet 1.1 and 1.1F Converting Celsius to Fahrenheit

This type of worksheet illustrates the ease with which cells may be filled by the Fill Down or Fill Right command under Edit. The set of operations illustrated by this example is representative of a process for copying and creating relative cell references, with the software intuitively changing the cell reference as you fill down or to the right. The process of filling cells where one or more of the referenced cells has an absolute address will be revealed later in this chapter. If you need assistance in producing these results, a step-by-step sequence for developing this worksheet is detailed in Operations List 1.1. The alphanumeric characters that are presented in bold type are the only characters that you will have to type into the cells. The remainder of the spreadsheet can be completed by using the Fill Down command under Edit. It is imperative that you leave cell A1 empty (what appears in this cell is the overflow from the right justified contents of cell B1) so that this worksheet may be used for a graphical presentation later. If you wish to improve the appearance of a worksheet, you can align the contents of a cell or a group of cells by using the Alignment option under Format on the menu bar. The Alignment option allows you to select General, Left, Center, Right, or Fill for aligning the contents of the cell or cells that you have selected by using the mouse or by using the shift key in conjunction with the arrow keys.

Operations List 1.1 Operations for Producing Worksheet 1.1

B1, enter **Celsius -> Fahrenheit**, then return
B1, Format, Alignment, Right, OK
A2, enter **0**, then return
A3, enter **=A2+10**, then return
A3, press and hold the mouse button, drag down to A12, release, Edit, Fill Down
B2, enter **=9/5*A2+32**, then return
B3, press and hold the mouse button, drag down to B12, release, Edit, Fill Down

Graphical Presentations

The worksheet is now complete and you are ready to investigate some of the remaining features of this tool. You may select cell A2 and enter any number you wish. It is suggested that you enter the number -50 and observe the results (you should notice the temperature value that is the same on both of these temperature scales). The presentation of this information in graphical format (a chart) may be quickly achieved by the operations that are listed in Operations List 1.1A. For this graph, enter a zero in cell A2 and your results will be identical to Chart 1.1, which is shown on the next page.

Operations List 1.1A Operations for Producing Chart 1.1

A1, press and hold the mouse button, drag down to A12 and across to B12, release, File, New, Chart, OK

Gallery, Line, 1, OK

Chart, Attach Text, Category Axis, OK, enter **Celsius**, then return

Chart, Select Chart, enter **Fahrenheit**, then return, place arrow on Fahrenheit and press the mouse button, drag the rectangle to a position above 250, release, move arrow to empty space and click

The graph should now be complete with a title for the chart, the category axis labeled, and the value axis labeled. Later, you may want to explore the various options for graphical presentations under Gallery on the menu bar. Appendix C has information on changing the size, moving, shrinking, and enlarging the chart window. The creation of this new chart (or a new worksheet) has the effect of "stacking it on top" of the previous documents. With the Window command on the menu bar, it is possible to switch back and forth between any documents that you have created. To return the worksheet to your view, select Window, and then 2 Sheet1. The Window command allows you to switch between different worksheets and charts. With the worksheet as the current window, enter in cell A2 the numeric value -50 and then use the Window followed by 1 Chart1 commands to activate a view of the graph. You should observe that the chart now represents the new values that have been calculated by the worksheet for an initial value of -50° C. Now select Window from the menu bar followed by Arrange All. You should observe a side-by-side presentation of both the worksheet and the chart. If you activate the worksheet by clicking on cell A2, you will be able to change this value in this cell at will and observe the corresponding changes in the graphical presentation. You should notice that the menu bar corresponds to the document that you have activated. It is possible to reduce the horizontal size of the worksheet by moving the borders and thereby gaining space so that the chart may be increased in size in the horizontal direction. This will allow room for the numerical scale values on the category axis of the chart. This technique of linking the worksheet and the chart is a very useful tool for exploring the consequences of varying different parameters for many of the chemical systems we will investigate in this book.

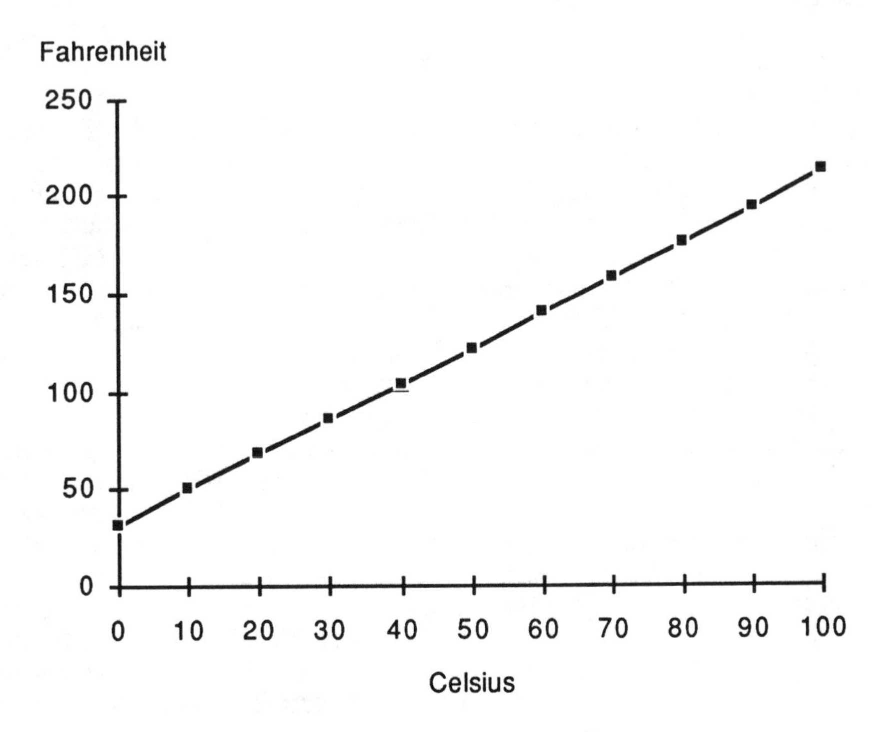

Chart 1.1 Converting Celsius to Fahrenheit

Viewing Formulas and Adjusting Cell Width

You can view the formula that you have placed in a given cell by observing the formula bar in the third line at the top of the worksheet. This requires that you highlight the cells one at a time by using either the mouse or the directional arrows. During the development of a worksheet, you may want to simultaneously observe the formulas in all of the cells. This can be accomplished by altering the view of the worksheet so that only the formulas and not the contents of all of the cells are shown. The following sequence of commands will produce the desired result.

Options, Display, Formulas, OK

This process may be repeated in order to return the view to that of the cell contents. In many cases the contents in a cell will not be visible because the width of the cell is too narrow. This can be changed by using the following sequence of commands.

Format, Column Width, "**numerical value**", OK

Absolute Cell References

Many applications require that several cells reference one or more cells repeatedly as an absolute address. It is possible to use the Fill Down or Fill Right command and refer to the same cell or absolute cell in all of the cells that you fill. There are two ways that this may be accomplished. The first way that we will use is the method of typing a ($) before the column and row coordinates in your formulas. This procedure will ensure that a reference to this cell will always be made when the Fill Down and Fill Right commands are used. If you place a ($) in front of the alphabetical character, then any operation that fills additional cells will hold the column constant and let the row vary. If you place a ($) in front of the numerical character, then the filling operations will hold the row constant and let the column vary. As an example, we will produce a worksheet that will convert the Celsius temperature to Kelvin. In all of the calculations a constant value of 273.15 has to be added to the Celsius temperature. When constants are needed for calculations that you intend to accomplish on a spreadsheet, convenience dictates that they be placed in the upper cells of the worksheet. You may wish to label the meaning of a constant by entering alphanumeric characters in neighboring cells. Worksheet 1.2 illustrates the use of a constant in cell A1; this value is needed in all of the cells in column B and it is referred to as $A$1 in all of the formulas. Cell B3 is the only cell that needs to have the contents typed into it; the remaining cells are filled by using the Fill Down command.

	A	B
1	273.15	
2	Celsius ->	Kelvin
3	-250	=A3+$A$1
4	=A3+50	=A4+$A$1
5	=A4+50	=A5+$A$1
6	=A5+50	=A6+$A$1
7	=A6+50	=A7+$A$1
8	=A7+50	=A8+$A$1
9	=A8+50	=A9+$A$1
10	=A9+50	=A10+$A$1
11	=A10+50	=A11+$A$1

Worksheet 1.2F Converting Celsius to Kelvin

After you have completed this worksheet, you may wish to use it to review the operations needed for graphing a function. You can change the value of the Celsius temperature in cell A3, display the data as a chart or graph, and link the chart to the worksheet with the Arrange All command under Window on the menu bar. Figure 1.1 illustrates the simultaneous viewing of a chart and a worksheet.

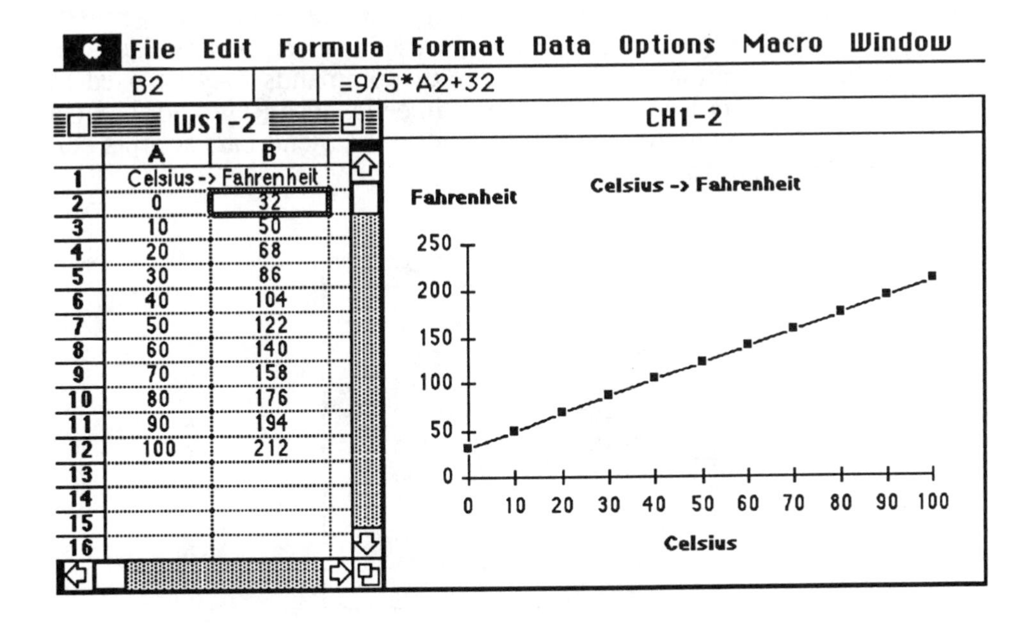

Figure 1.1 Chart 1.2 and Worksheet 1.2 Linked

Absolute Cell References by Defining Names

Excel provides a second method for establishing an absolute cell reference with the formula define name commands. You may create a name (up to 255 characters) for a cell and then refer to this name in the formulas that you place in other cells. To name a cell you must first select it and then choose the Define Name command on the Formula menu. Using Worksheet 1.2 as an example, select cell A1 and proceed to name this cell Ab or Absolute. The appropriate sequence of commands is

A1, Formula, Define Name,
In the dialog box under Name:, **Ab**, OK

B3, **=A3+Ab**, then return
select cell range B3 -> B11
Edit, Fill Down

You will see that Worksheet 1.3 is identical to Worksheet 1.2 in appearance and function, but the underlying formulas are different, with Worksheet 1.2 using $A$1 as an absolute cell reference and Worksheet 1.3 using the define name command to reference cell A1 in an absolute manner by the name Ab or Absolute. Either technique will serve this purpose and both will be used in this book. The formula define name command allows the input of formulas and equations in a symbolic form that is more familiar to chemists. The Formula, Define Name sequence may be used to create a name for a cell range, value, or formula. This sequence of commands may be used to define values or formulas that do not exist in cells. The Refers to: box (in the Define Name command under Formula on the menu bar) accepts cell references, values, or formulas.

	A	B
1	**273.15**	
2	**Celsius - >**	**Kelvin**
3	**-250**	**=A3+$A$1**
4	**=A3+50**	=A4+$A$1
5	=A4+50	=A5+$A$1
6	=A5+50	=A6+$A$1
7	=A6+50	=A7+$A$1
8	=A7+50	=A8+$A$1
9	=A8+50	=A9+$A$1
10	=A9+50	=A10+$A$1
11	=A10+50	=A11+$A$1

Worksheet 1.3F Cell Reference by a Defined Name

Problem Solving With Spreadsheets

The last example in this chapter will demonstrate the ease, speed, and versatility that spreadsheets bring to the task of problem solving. The example will involve the calculation of the mass in kilograms of gold spheres that have diameters ranging from 5 cm to 13 cm. The technique employed places all of the constants needed for the mass calculations in cell A1 in the form of a mathematical formula. This allows you to view, correct, and change the mass units with ease because the underlying formula is presented in an uncalculated form. This is a useful technique, as it allows you and others to later observe your method of solving the problem. The underlying values and formulas are shown in Worksheet 1.4. There are several problems at the end of this chapter that will assist you in developing these problem solving skills. In addition, chemistry texts will have many examples of problems that may be solved by these spreadsheet techniques.

	A	B
1	=(4/3)*3.14*19.3/1000	**Au**
2	diameter	
3		kg
4	5	=$A$1*(A4/2)^3
5	=A4+1	=$A$1*(A5/2)^3
6	=A5+1	=$A$1*(A6/2)^3
7	=A6+1	=$A$1*(A7/2)^3
8	=A7+1	=$A$1*(A8/2)^3
9	=A8+1	=$A$1*(A9/2)^3
10	=A9+1	=$A$1*(A10/2)^3
11	=A10+1	=$A$1*(A11/2)^3
12	=A11+1	=$A$1*(A12/2)^3

Worksheet 1.4F Mass of Au Spheres

The sphere volume is calculated from the equation, $V = (4/3) \cdot \pi \cdot r^3$. The value of $(4/3) \cdot \pi$ will be used in each mass calculation. The density of Au is 19.3 g/cm^3 and this is needed in each calculation. The answer at this point would be in grams, but this example will be calculated in kilograms and thus the division by 1000. Others can view your work by checking the underlying formula that is presented in A1, and the choice of the units for the final answer may be quickly and conveniently changed. The number of significant digits can be controlled by choosing the Number command on the Format menu. The resulting data and calculations may be graphed, and this graph and the worksheet may be linked by their simultaneous presentations on the monitor screen. The functional relationship between the mass of gold spheres and their diameters is illustrated in Chart 1.4.

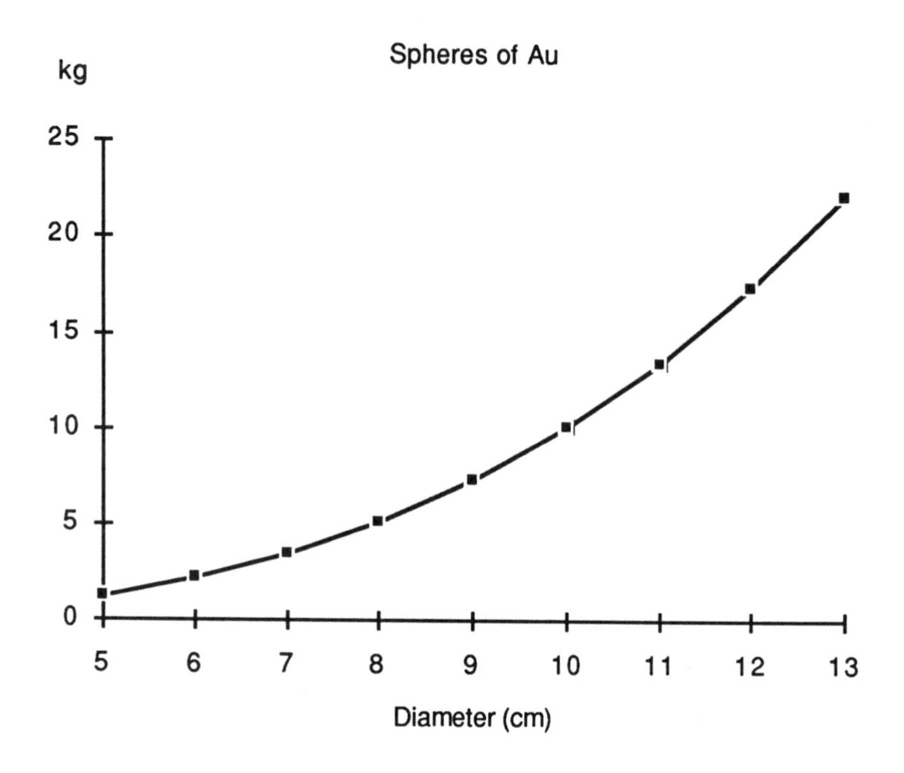

Chart 1.4 Mass of Gold Spheres

Problem solving with spreadsheet software offers many quick and convenient options. Modern spreadsheets are friendly and as easy to use as contemporary calculators. The quick presentation of data in a tabular and/or graphical form in conjunction with the option of quickly changing values, formulas, and results are very useful functions for scientists and engineers.

The Edit options allow you to Cut, Copy and Paste, and Clear cells or groups of cells. The Edit menu also contains the Delete and Insert commands, which will insert or delete a line if the row coordinate (numerical) is selected. If you select a column coordinate (alphabetical), this command will insert or delete a column. The Format options allow you to change the presentation of the Number, the Alignment within a cell, the Font, and the Column Width. The Window options allow you to switch to the desired worksheet, chart, or macro sheet.

A copy of your work may be produced with the Print command on the File menu. Control of the presentation is regulated by Page Setup and

Printer Setup under the File menu. This will vary slightly depending upon whether you are using an MS-DOS version of Excel for IBM compatible computers or an Excel version for Macintosh computers.

Problems

1. Develop a worksheet for converting Fahrenheit to Rankin and graph the results. The scales are converted by °R = °F + 459.67.

2. Develop a worksheet for converting Kelvin to Rankin. Graph this data and your results.

3. Develop a worksheet for converting human weights from 100 lbs to 200 lbs to kilograms in intervals of 10 lbs. Graph the results.

4. Develop a worksheet for converting heights from 5 ft to 6.5 ft to meters in 2-in. intervals. Graph your data.

5. Develop a worksheet for converting gals of gasoline to lbs for values of 10 gal to 20 gal in 1-gal intervals. D(gasoline) = 0.65 g/mL, 1.0 quart (US) = 946.35 cm^3. Graph your data. See the next page.

6. Develop a worksheet for calculating the mass of spheres of Au that range in size from 1 in. to 10 in. in diameter. Use 1-in. intervals. Graph this data. D(Au) = 19.3 g/cm^3.

7. Mountain heights on the North American Continent range from 10,000 ft to 20,000 ft. Convert these heights to meters and graph this data.

8. Convert engine sizes in cubic inches to liters. Use a range of 100 $in.^3$ to 300 $in.^3$ with 20 $in.^3$ intervals. Your conversion factor should have this form: **=2.54^3*0.001**. Graph your results.

9. The density of air at 25°C and atmospheric pressure is 1.19 g/L. Design a worksheet that will calculate the weight (pounds) of air contained in rooms of different sizes (dimensions in feet). Enter the density in an individual cell so that it may be changed for different conditions and substances. Place the conversion factor in a separate cell and then allow individual cells for the user to enter the length, width, and height of the room. As a test of your sheet, enter 1000 g/L and calculate the weight of a cubic foot of water. You should have 62.4 lb/ft^3. An answer is presented on the next page.

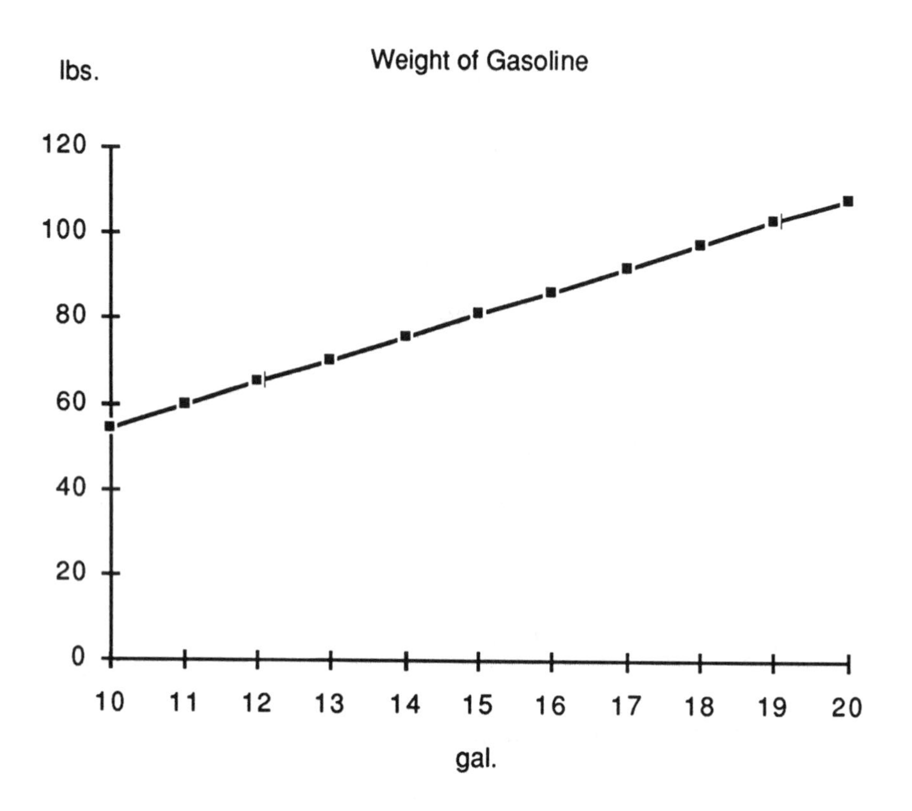

Chart 1.5 A Sample Chart For Problem 5

	A	B
1	**Weight of Air in a Room**	
2		
3	1.19	Air (g/L)
4	=A3/453.6*0.001*2.54^3*12^3	Conversion Factor
5		
6	10	<- Length (ft)
7	10	<- Width (ft)
8	10	<- Heigth (ft)
9	=A6*A7*A8	Volume (cu ft)
10	**=A4*A9**	Weight (pounds)

Worksheet 1.9 A Worksheet For Problem 9

2

Density and Conditional Tests

Conditional tests are equations that compare numbers, formulas, or labels. A very important function that can be used to produce branching or alternative calculations on a spreadsheet is the IF function. This function has a form:

=IF(*conditional test, true value, false value*)

It is possible to use other functions as the arguments, to use text as arguments (text characters must be surrounded by double quotes), and to nest IF functions as arguments. Three additional functions let you develop compound conditional tests: AND, OR, and NOT. These functions can be used in conjunction with the logical operators =, >, <, >=, <=, and <>. Arguments for AND, OR, and NOT can be conditional tests, arrays, or references to cells. (BLB Chap. 1)

Measuring Density

Density is a characteristic physical property of all matter. It is useful for the identification of materials, for the separation of minerals, for calculating mass in commerce and architecture, and for calculating the buoyancy of objects. An experimental method for finding the density of a solid involves measuring the mass of the object in air and in a liquid such as water. The difference between the mass in air and in water divided by the density of water is the volume of water displaced or the volume of the solid. When the measurement is made at 20°C, the density of water is 0.9982 g/mL, and the density of the solid is given by

$$D = \frac{m_{air}}{m_{air} - m_{water}} \times 0.9982$$

An example of a worksheet that makes these types of calculations is presented as Worksheet 2.1F. The underlying formulas are shown so that you may produce your own spreadsheet. You should notice the two conditional tests that occupy cells B7 and B8 that accommodate the solid having a density equal to or less than the liquid. The conditional statement in cell B7 allows for a situation where the solid and the liquid have the same density. The statement in cell B8 alerts the user to the reason for a calculated negative density.

	A	B	C
1	**Density -**	by measuring the mass in air and in a liquid	
2			
3	m (air)	17.43	**<- input**
4	m (liquid)	10.99	**<- input**
5	D (liquid)	0.9982	
6			
7	**Density**	=IF(B3=B4,B5,B5*B3/(B3-B4))	g/mL
8		=IF(B4>B3,"Solid less dense than liquid"," ")	

Worksheet 2.1F Measuring the Density of a Solid

Densities for solids, liquids, and gases at 20°C are given in Table 2.1. The densities of the gases are given for a pressure of one atmosphere.

Table 2.1 Density at 20°C in g/cm^3

Au (*s*)	19.3	Hg (*l*)	13.59
Pb (*s*)	11.3	H2O (*l*)	0.9982
Ag (*s*)	10.5	CH_3OH (*l*)	0.79
Ni (*s*)	8.90	gasoline (*l*)	0.64
Fe (*s*)	7.86	air (*g*)	0.001197
Al (*s*)	2.70	H_2 (*g*)	0.000083
Mg (*s*)	1.74	He (*g*)	0.000166
H_2O (*s*)	0.917	Kr (*g*)	0.003484

Lift and Apparent Mass

The lift or apparent loss of mass when an object is immersed in a fluid (a liquid or gas) can be calculated. The lift exerted on an object is equal to the mass of fluid displaced by the object. This lift, when subtracted from the mass of the object in air, yields the apparent mass or the mass of the object when it is totally submerged in the specific fluid. In order to broaden the scope of our calculations, some additional densities are listed in Table 2.2.

Table 2.2 Density at 20°C in g/cm^3

Topaz (s)	3.49	C_6H_6 (l)	0.894
Epsonite (s)	1.68	CCl_4 (l)	1.595
Agate (s)	2.65	$C_2H_2Br_4$ (l)	2.964
NaCl (s)	2.16	CH_2I_2 (l)	3.325

The liquids listed in Table 2.2 may be used to find the density of crystals that are so small that conventional methods will not measure their volume or mass. The density of these small crystals may be accurately measured by testing their buoyancy in liquids of different density. Various ratios of CCl_4 and CH_2I_2 may be used to prepare a liquid with any desired density between 1.595 and 3.325. For liquids of this type, the volumes are essentially additive. Mixtures of these two liquids can be used to separate four of the solids, topaz, epsonite, agate, and sodium chloride, given in Table 2.2. By adjusting the ratio of these two liquids to produce liquids of different densities, the densities of the four solids can be determined. As an example, a 50/50 (vol/vol) mixture would have a density of 2.460, and it would separate agate and topaz from sodium chloride and epsonite, as the former two solids would both sink and the latter two solids would both float in this fluid. Worksheet 2.2 calculates the lift provided by a fluid that surrounds an immersed solid, the measured apparent mass of the solid immersed in the liquid, and its density in air. Three experimental values are necessary in order for these calculations: the mass and volume of the solid, and the density of the liquid in which the solid is immersed.

	A	B	C
1	**Lift and apparent mass**		
2			
3	V (mL)	2	<- input
4	D (liquid)	1.595	<- input
5	m (g)	4.326	<- input
6			
7	lift	**3.19**	g
8	mass in liq	**1.136**	g
9	D (object) air	**2.163**	g/mL
10			
11	mass vacuum	4.328	g
12	absol error	0.0024	g
13	relative error	0.553	ppt
14	D (obj) vacuum	**2.164**	g/mL

Worksheet 2.2 Lift and Apparent Mass

Air may be considered a fluid, and Worksheet 2.2 is designed to calculate the mass of the solid in a vacuum as well as its mass in the specified liquid. As noted before, the lift produced by the specified liquid is calculated and the result presented in conjunction with the density of the solid in air. The mass of the object in a vacuum is calculated by adding to the mass measured in air the mass of the air displaced by the solid. The difference between the mass of the solid in air and in a vacuum is calculated and presented as the absolute error. The relative error of a mass measurement in air as compared to the true mass that would be determined in a vacuum is also presented. The last calculation presents the true density of the solid. Sodium chloride is used as the solid and the liquid is CCl_4.

The formulas for Worksheet 2.2 are presented in the following view of Worksheet 2.2F. The mass of the object in a vacuum will be larger than the mass in air due to the lack of a "fluid" to provide the lift that is inherent with any measurement made with air present. The volume of the object has to be multiplied by the density of air (dry air has been used in these examples) and then added to the measured mass of the object in air.

	A	B	C
1	**Lift and**	**apparent mass**	
2			
3	V (mL)	2	<- input
4	D (liquid)	1.595	<- input
5	m (g)	4.326	<- input
6			
7	lift	**=B3*B4**	g
8	mass in liq	**=B5-B7**	g
9	D (object) air	**=B5/B3**	g/mL
10		=IF(B8<0,"Solid less dense than liquid"," ")	
11	mass vacuum	=B5+0.001197*B3	g
12	absol error	=B11-B5	g
13	relative error	=1000*B12/B11	ppt
14	D (obj) vacuum	**=B11/B3**	g/mL

Worksheet 2.2F Lift and Apparent Mass

Worksheet 2.2 may be used to calculate the lift capacity of a balloon filled with a gas such as hydrogen or helium, that are "lighter" than air. An example of a balloon with a diameter of 60.0 cm which is filled with hydrogen is illustrated as Worksheet 2.3. The mass of the balloon is part of the total load that this hydrogen-filled balloon can lift. The volume is approximately 113,000 cm^3, and the mass of the hydrogen in the balloon is equal to the volume of the balloon multiplied by the density of hydrogen.

	A	B	C	D
1	**Lift and apparent mass**			
2				
3	V (mL)	113000	**<- input**	
4	D (liquid)	0.001197	**<- input**	
5	m (g)	9.47	**<- input**	
6				
7	lift	**135.26**	g	
8	mass in liq	**-125.79**	g	
9	D (object) air	**0.0000838**	g/mL	
10		Solid less dense than liquid		
11	mass vacuum	144.73	g	
12	absol error	135.26	g	
13	relative error	934.57	ppt	
14	D (obj) vacuum	**0.0012808**	g/mL	

Worksheet 2.3 Balloon Filled With Hydrogen

For the density of the liquid, the density of air is used, 0.001197 g/cm^3. The lift presented on the spreadsheet is the total lift provided by the fluid, "air," and does not include the mass of the hydrogen. The apparent mass of hydrogen (the solid) in air (a liquid) is in this case the total mass that the balloon can carry. The computed density is that of hydrogen. The mass in a vacuum is incorrect for this example because the object density is less than the fluid density. The computed results for solids that are less dense than the fluid in which they are immersed are incorrect, because they float and are not completely surrounded by the fluid. This spreadsheet may be used to investigate the lift provided by other gases such as helium and methane.

Correcting the Measured Density

The previous example for calculating the mass of the object in air and in a vacuum required a knowledge of the volume of the object. This volume measurement can be obtained by finding the mass of the object in air and its apparent mass as measured when it is immersed in a fluid of known density. Worksheet 2.4 shows a view of the underlying formulas that are used for calculating the density of the solid in air and in a vacuum when the mass of the solid is measured in both air and a liquid of known density. This spreadsheet does not calculate the corresponding values for the solid in a vacuum when the density of the solid and the liquid in which it is immersed are equal. In this case the solid will float, and it is not possible to measure its volume, which is necessary in order to calculate the values for the object under the conditions of a vacuum.

	A	B	C
1	**Density -**	by measuring the mass in air and in a liquid	
2			
3	m (air)	17.43	**<- Input**
4	m (liquid)	10.99	**<- Input**
5	D (liquid)	0.9982	g/mL
6			
7	**Density**	**=IF(B3=B4,B5,B5*B3/(B3-B4))**	g/mL
8		=IF(B4>B3,"Solid less dense than liquid"," ")	
9	m (vacuum)	=B3+0.001197*(B3-B4)/B5	g
10	absol error	=B9-B3	g
11	relative error	=1000*B10/B9	ppt
12	D (vacuum)	**=B5*B9/(B3-B4)**	g/mL
13		=IF(B3=B4,"vac calculations incorrect"," ")	

Worksheet 2.4 Density Measured in Air and in a Vacuum

Problems

1. Calculate the apparent mass of a 1.00 cm^3 crystal of topaz in air, water, CCl_4, and in CH_2I_2.

2. Calculate the relative error produced when the mass of a solid is measured in air as opposed to being measured in a vacuum. For these calculations use a volume of 500 cm^3 and 2500 cm^3 for five solids of different densities. Use Au, Pb, Fe, Al, and Mg as the solids. How does the relative error vary as a result of a difference in density? Comment on the effect that volume of a given solid has on the relative error.

3. Calculate the apparent mass of a 10.4 cm^3 crystal of epsonite in air, water, C_6H_6, and CCl_4.

4. Calculate the apparent mass of a 5.62 cm^3 crystal of agate in water, C_6H_6, CCl_4, $C_2H_2Br_4$, and CH_2I_2.

5. Calculate the apparent mass of a 10.0 cm^3 block of magnesium when immersed in water, C_6H_6, CCl_4, and CH_2I_2.

6. Calculate the relative error when a 100 cm^3 block of magnesium is "weighed" in air.

7. Pyrite has a density of 5.018 g/mL. Compute the apparent mass of a 1.000 cm^3 crystal of this mineral when it is immersed in water. Use the computed apparent mass and the mass of 1.000 cm^3 of this mineral in air to recalculate the density of pyrite in a vacuum. Calculate the apparent mass of a 4.890 cm^3 crystal when immersed in C_6H_6, CCl_4, and $C_2H_2Br_4$.

8. Calculate the lift provided by a 100 cm (diameter) weather balloon that is filled with helium. You may choose to calculate the volume by a sequence of arithmetic steps in the cell requiring the volume.

9. Imagine a cube that has 10 cm sides as outside to outside dimensions. This cube has a bottom and four sides constructed of a material that is 1 cm thick. Design a spreadsheet that will calculate the maximum density a material used for constructing the walls can have and still have a cube that will float in sea water (1.025 g/mL). You should find that the total volume of the walls is 424 cm^3. Your calculated results for the "empty volume" which contains air should be 576 cm^3. Use your spreadsheet to calculate the maximum density that the material may have if the cube is 100 cm on each side as an outside to outside measurement. The walls of the new cube are still 1 cm thick.

	A	B	C	D	E	F
1	**Lift and apparent mass**					
2						
3	V (mL)	1	<- input	4.89	4.89	4.89
4	D (liquid)	0.9982	<- input	0.894	1.595	2.964
5	m (g)	5.018	<- input	24.538	24.538	24.538
6						
7	lift	**0.9982**	g	**4.372**	**7.800**	**14.494**
8	mass in liq	**4.0198**	g	**20.166**	**16.738**	**10.044**
9	D (object) air	**5.018**	g/mL	**5.018**	**5.018**	**5.018**
10						
11	mass vacuum	5.019	g	24.544	24.544	24.544
12	absol error	0.0012	g	0.0059	0.0059	0.0059
13	relative error	0.238	ppt	0.238	0.238	0.238
14	D (obj) vacuum	**5.019**	g/mL	**5.019**	**5.019**	**5.019**
15						
16		in H20		in C6H6	in CCl4	C2H2Br4

Worksheet 2.5 A Worksheet For Problem 7

3

Treating Data

A dramatic forte of spreadsheets is the ease and convenience they provide for the treatment of experimental data. Our first consideration will be the experimentally determined mass of the different isotopes of the chemical elements of the periodic table. There are some 2600 known isotopes of the 108 elements presented on the modern periodic table. Various tabulations of the isotopes contain data on their mass, the percent abundance, stability, and decay mode if the isotope is a radioactive one. With this data an average atomic mass for a specific element can be calculated. The average atomic mass or the atomic weight of the different elements establishes the basis for quantitative measurements and calculations in chemistry. Next, we will use the atomic weight to calculate a total mass or molecular weight for a few selected compounds. These calculations lead to a consideration of the mass percentage of a specific element in selected compounds. Our final consideration will be the statistical treatment of data that results from laboratory measurements. We will examine the Gaussian distribution and the ease with which statistical calculations can be made with the assistance of spreadsheets. The availability of defined functions with the spreadsheet software will be highlighted. (BLB Chap. 3)

Atomic Mass

The determination of the average atomic mass for the elements of the periodic table is fundamental to any quantitative measurements made by chemists. Tables of the isotopes list the relative abundance and atomic mass for specific elements. The sum of the products of the decimal fraction that represents the relative abundance and the individual masses of the isotopes of an element is the average atomic mass for that element. This value is known to chemists as the atomic mass or atomic weight. Worksheet 3.1 illustrates the calculation of this value for the element mercury. Data on the atomic mass of isotopes can be found in *CHEMISTRY The Central Science*, 5th Edition, by Brown, LeMay, and Bursten, or in *Chart of the*

Nuclides, 13th Edition, by Walker, Miller, and Feiner, published by General Electric, San Jose, CA.

	A	B	C	D
1	Atomic weight - the average atomic mass			
2	Isotopes	Hg	<-element	
3		% abund	Mass	
4	196	0.140%	195.965	0.274
5	198	10.039%	197.967	19.874
6	199	16.830%	198.967	33.486
7	200	23.120%	199.968	46.233
8	201	13.230%	200.97	26.588
9	202	29.790%	201.97	60.167
10	204	6.850%	203.973	13.972
11				
12		The atomic weight is		**200.59**

Worksheet 3.1 Atomic Mass of Mercury

This is the accepted value for the atomic weight of mercury. This value may be used in all calculations that involve mercury with the assumption that the isotopic distribution is a constant on this planet. There are some significant variations in the isotopic distribution of some elements, but in most cases the assumption of a constant isotopic distribution is valid. The deviation in isotope distribution is a result of artificial isotopic enrichment or separation, artificial nuclear reaction, localized geological occurrence because of the radiogenic source, or samples of extraterrestrial origin.

Formula List 3.1 Formulas for Worksheet 3.1

Cell D4	=B4*C4	=abundance*amu (isotopes)
Cell D12	=SUM(D4:D10)	=atomic mass

The remainder of the formulas in column D are copied from cell D4 to cells D5 through D10 by using the Fill Down command under Edit on the menu bar. The precise atomic mass of each isotope of the element is multiplied by the decimal value of its natural abundance in each row of the spreadsheet. The atomic mass or atomic weight of an element is then computed as the sum of these product values for all of the isotopes known to exist on this planet of the specified element. The atomic mass represents the mass in grams of 6.0221×10^{23} atoms of the chosen element.

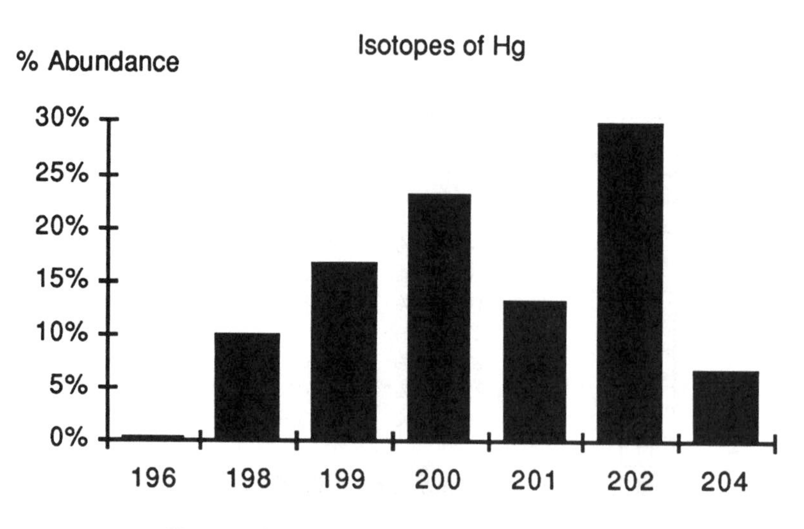

Chart 3.1 Isotopic Distribution of Mercury

Data for the four stable isotopes of chromium are presented as the underlying formulas in Worksheet 3.2F. The percent abundance data may be entered as a decimal value in column C and the presentation of the values formatted so that the percent values appear. In order to accomplish this, you first select cells C4 through C8, then choose Format, Number, one click on 0.00%, and then move the cursor to the Format dialog box in order to add zeros so that it reads 0.0000%, OK. The addition of a sequence of cells may be accomplished by using the formula, **=SUM(D4:D7)**, or by the normal notation of adding each individual cell as a string of sums, **=D4+D5+D6+D7**. For convenience, the two different forms of summation notation may be mixed in any combination or sequence.

	A	B	C	D
1	Atomic	weight - the	average	atomic mass
2	Isotopes	Cr	<-element	
3		% abund	Mass	
4	50	0.0435	49.9461	=B4*C4
5	52	0.8379	51.9405	=B5*C5
6	53	0.095	52.9407	=B6*C6
7	54	0.0236	53.9398	=B7*C7
8				
9		The atomic	weight is	=SUM(D4:D7)

Worksheet 3.2F Isotopes and Atomic Weight of Cr

An alternative method for calculating the sum of the products of two rows or columns is available with Microsoft's Excel. Cell C9 contains an expression that is an example of an array formula. The evaluation of arrays is specific to Excel, and other spreadsheet software may or may not evaluate expressions of this form. In order to enter an array, it is necessary for you to hold down the Command key when you press the Enter key or click the enter box after you have placed the formula in cell C9. The formula in C9

=SUM(B4:B7*C4:C7)

calculates the atomic mass. If the formula is correctly entered, it will appear in the formula bar enclosed in brackets. These brackets are not typed around the formula; they appear when the array is correctly entered by holding down the Command key when the Enter key is pressed.

{=SUM(B4:B7*C4:C7)}

This design allows for the complete removal of column D. An example of the results of this design is illustrated as Worksheet 3.3. The array formulas that are available with Excel can be used for any mathematical operation that requires a row-by-row and column-by-column procedure.

	A	B	C
1	Atomic mass		
2	Isotopes	Cr	<-element
3		Mass	% abund
4	50	49.9461	4.35%
5	52	51.9405	83.79%
6	53	52.9407	9.50%
7	54	53.9398	2.36%
8			
9		Atomic mass =	51.996

Worksheet 3.3 Atomic Mass Using an Array

Percent Mass Composition

The molecular weight or total mass of a molecular substance is equal to the sum of the products of the atomic weights of the individual elements multiplied by the number of atoms of that element in the molecular formula. The percent composition by mass for an element is equal to the atomic weight of the element multiplied by the number of atoms of that element in the molecular formula divided by the molecular weight and then multiplied by 100 to convert to percent. Worksheet 3.4 calculates the molecular weight of each compound and the percent of each element by mass in a specified compound. The following three compounds are used in this spreadsheet: $(NH_4)_2SO_4$, $CO(NH_2)_2$, and NH_4NO_3. Ammonium nitrate is the obvious choice for the compound with the maximum amount of nitrogen when based on the percent by weight. More detailed calculations involving an extension of the concept of molecular weight will be presented in the next chapter. This application emphasizes the rapid calculation of the percent composition by mass for compounds that all contain an element of interest.

	B	C	D	E	F	G	H
1	**% N by Mass**						
2							
3		(NH4)2SO4		CO(NH2)2		NH4NO3	
4	1.0079	8.06	6.1%	4.032	6.7%	4.0316	5.0%
5	14.0067	28.01	**21.2%**	28.013	**46.6%**	28.0134	**35.0%**
6	12.011			12.011	20.0%		
7	15.9994	64.00	48.4%	15.999	26.6%	47.9982	60.0%
8	32.06	32.06	24.3%				
9							
10	mol. wt.	132.13		60.055		80.043	

Worksheet 3.4 Percent Mass Composition

	C	D	E	F	G	H
4	=8*B4	=C4/C10	=4*B4	=E4/E10	=4*B4	=G4/G10
5	=2*B5	**=C5/C10**	=2*B5	**=E5/E10**	=2*B5	**=G5/G10**
6			=B6	=E6/E10		
7	=4*B7	=C7/C10	=B7	=E7/E10	=3*B7	=G7/G10
8	=B8	=C8/C10				
9						
10	=SUM(C4:C8)		=SUM(E4:E8)		=SUM(G4:G8)	

Worksheet 3.4F Percent Mass Composition

Statistics

The majority of the values used by chemists are obtained as measurements in the laboratory. All these measurements are subject to error, regardless of the source. Experimental errors are an inherent characteristic of measurements whether they are made with simple volumetric glassware or with the most sophisticated microprocessor controlled instrument. The rational interpretation of data requires an analysis of the variation intrinsic in any series of measurements. Systematic or determinate error can at least in principle be corrected by careful experimental techniques. In practice this principle is much easier articulated than accomplished. Systematic error is nonrandom, and it is therefore not amenable to the mathematical treatment developed for random or indeterminate error.

Several different physical phenomena are governed by chance, as well as sets of data with random error that have been obtained from experiment measurements. All of these events can be mathematically described by a Gaussian distribution. Our concern is the analysis of random or indeterminate error by statistical analysis which implies the application of a Gaussian distribution. The mathematical function that describes this random behavior is an exponential function that is symmetrical about μ, the unattainable true value (assumes an infinite set of data).

$$y = \frac{1}{\sigma\sqrt{2\pi}} e^{-(x-\mu)^2/2\sigma^2}$$

In this function, e is the base of the natural logarithm, 2.718, μ is the population mean, and σ is the population standard deviation for an infinite set of data. We can never measure μ and σ, but as the number of measurements increase, the values of $\bar{x}$ and s approach μ and σ. The quantity $\bar{x}$ is the mean value of several measurements of x, and the mathematical definition is given on page 28. For a finite number of measurements, s is the parameter that corresponds to σ for an infinite set of data. The mathematical definition of s is given on page 29.

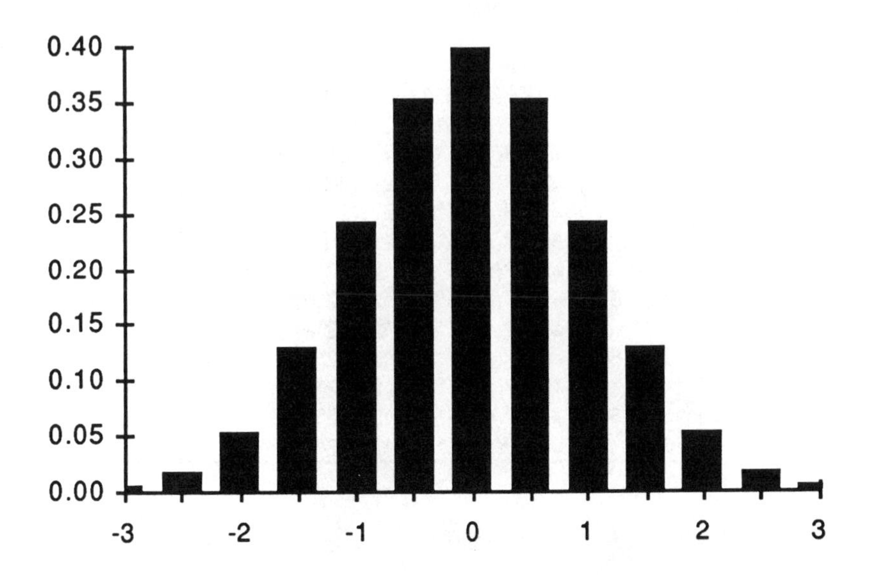

Chart 3.5 A Gaussian Distribution

Chart 3.5 illustrates the general shape of this function when $\mu = 0$ and $\sigma = 1$. Worksheet 3.5 which produced this chart is shown at the bottom of this page. The relative probability of any particular value of x, which is experimentally measured, is proportional to the value of y or the value on the vertical scale. For a finite number of measurements, that graph appears as a sequence of bars. For an infinite number of measurements, the function produces a smooth envelope around the bars. The general shape of this function is the basis for the analysis of random error.

The most probable value of x in a series of measurements is that value of x that produces the maximum value of the Gaussian function. The probability of a measured value of x lying within a range is proportional to the area within that range. With reference to Chart 3.5, the area under the curve from an x value of -1 to 1 is .6826 or 68.26% of the total area under this curve. A set of measurements with their intrinsic random error will have 68.26% of the measured values within a range of ± 1 as presented with Chart 3.5 or within a range of ± s, which is known as one standard deviation. Your analytical considerations of this graph should not overlook the concept that the total area under the distribution function represents the total probability for all of the measurements or values. The sum of the probabilities of all the measurements must be one, hence the area under the curve has to be equal to unity when the range of x is from $-\infty$ to $+\infty$ for the area calculation. Similar considerations show that ± 2s represents 95.46% of the total area and ± 3s represents 99.74%.

	A	B
1	A	**Gaussian Curve**
2	X axis	
3		Y axis
4	-3	=EXP((-0.5*A4^2))/(2*PI())^0.5
5	-2.5	=EXP((-0.5*A5^2))/(2*PI())^0.5
6	-2	=EXP((-0.5*A6^2))/(2*PI())^0.5
7	-1.5	=EXP((-0.5*A7^2))/(2*PI())^0.5
8	-1	=EXP((-0.5*A8^2))/(2*PI())^0.5
9	-0.5	=EXP((-0.5*A9^2))/(2*PI())^0.5
10	0	=EXP((-0.5*A10^2))/(2*PI())^0.5
11	0.5	=EXP((-0.5*A11^2))/(2*PI())^0.5
12	1	=EXP((-0.5*A12^2))/(2*PI())^0.5
13	1.5	=EXP((-0.5*A13^2))/(2*PI())^0.5
14	2	=EXP((-0.5*A14^2))/(2*PI())^0.5
15	2.5	=EXP((-0.5*A15^2))/(2*PI())^0.5
16	3	=EXP((-0.5*A16^2))/(2*PI())^0.5

Worksheet 3.5 Calculating a Gaussian Distribution

Notice that cells A4 to A16 contain the values of x for which a calculation of the y values or values of the Gaussian function are desired. The Gaussian function is entered in cells B4 to B16 with the value of x being supplied by the corresponding value in column A. These calculated values are then displayed by the spreadsheet as a bar graph in Chart 3.5. The display format for the abscissa has been altered by using the following sequence of commands. Use the mouse to position the cursor on the category axis (horizontal axis) and click. This should produce two small square boxes at the origin and the terminal end of the abscissa (marking of the axis is available for either axis). When either axis is selected in this manner, the Scale command under Format is available. This selection allows the operator to make adjustments in the presentation of the axis scale, Patterns, and Font for the axis which is selected. For Chart 3.5 the following adjustments have been made. Under Scale, which in this case selects the Category Axis Scale, the dialog box for "Number of Categories Between Tick Labels:" has been changed from 1 to 2. The tick in the dialog box for "Value Axis Crosses Between Categories" has been removed. Then, with the additional selection of "Patterns...", the tick mark in the Minor box under Tick Mark Type has been changed from Cross to Invisible. With these and other options for changing the scale of a graph, the user has a much larger choice of the format of the graphical presentation.

The use of mathematical functions such as EXP and PI() in this spreadsheet is typical of the calculation power available with this software. The next section of this chapter will present the necessary details of the operational techniques that are needed in order to fully utilize the mathematical functions.

Mathematical Functions

Spreadsheet software such as Excel and Lotus 1-2-3 have very long lists of available mathematical functions. You will find the Paste Function command under Formula on the menu bar of Excel an excellent tool for assisting you in both selecting and remembering the syntax of a particular function. First, select the specific cell in which you wish to enter the mathematical function and then proceed to use the Paste Function command. After you have opened the Paste Function box, the mouse can be used to scroll through and select the function that you want. An alternative method allows you to select the first letter of the compressed notation of the function as a mean of scrolling through the selections and then pressing the Enter key when you have the function you want. After the function is pasted into the formula bar, the blinking cursor is positioned between the parentheses that will contain the arguments. With many of the functions the form of the arguments is obvious, but with others it is not readily apparent. Excel has a clever option that assists you in placing within the parentheses of the function the arguments in their correct form. Before you enter the mathematical function in the formula bar when you have the Paste Function box open, you click or place a tick in the Paste Arguments box. Excel will then prompt you with the form of the argument. These prompts disappear when you enter the data or cells that you need as arguments within the parentheses of the mathematical function.

Further Statistical Considerations

The mean value, $\bar{x}$, of several measurements of x which are collectively known as x_i (Excel refers to the mean value as the average function) is defined by the following equation:

$$\bar{x} = \frac{\sum_{i=1}^{n} x_i}{n}$$

A set of data with intrinsic random errors can be described by the symmetrical Gaussian distribution. The value of $\bar{x}$ approaches μ and is the most probable value of x. The most probable value of x is the maximum value of the Gaussian function.

The standard deviation, s, is a measure of how closely the data is bunched about the mean or average value. For a finite set of data, the standard deviation can be calculated from the following equation.

$$s = \sqrt{\frac{\sum_{i}^{n} (x_i - \bar{x})^2}{n - 1}}$$

This is the same standard deviation that was described on page 24 in relationship to Chart 3.5. The probability of measuring a value of x in a specified range is proportional to the area within that range as described by a Gaussian distribution. As previously discussed, 68.26% of the total area under the curve is within the range described by $\bar{x} \pm s$. In the idealized situation where a Gaussian distribution governs a set of data with only random error, the probability of a measurement being within a range of $\bar{x} \pm 2s$ is 95.46%, and within a range of $\bar{x} \pm 3s$ is 99.74%.

The relative probability of making a particular measurement of x is proportional to the value of the Gaussian function (the y-axis value) for the specified value of x. Worksheet 3.6 is a repeat of the spreadsheet that was used to produce Chart 3.5 with the addition of three columns that calculate the relative probability of making the specified measurements.

	A	B	C	D	E
1	A Gaussian Curve				
2	X axis		Relative Probability		
3		Y axis			
4	-3	0.0044	1.0	0.011	
5	-2.5	0.0175	4.0	0.044	
6	-2	0.0540	12.2	0.135	1.0
7	-1.5	0.1295	29.2	0.325	2.4
8	-1	0.2420	54.6	0.607	4.5
9	-0.5	0.3521	79.4	0.882	6.5
10	0	0.3989	90.0	1.000	7.4
11	0.5	0.3521	79.4	0.882	6.5
12	1	0.2420	54.6	0.607	4.5
13	1.5	0.1295	29.2	0.325	2.4
14	2	0.0540	12.2	0.135	1.0
15	2.5	0.0175	4.0	0.044	
16	3	0.0044	1.0	0.011	

Worksheet 3.6 Relative Probability of Measurements

If the relative probability of making a measurement of -3 or 3 for x is given a value of one, then the relative probability of making a measurement of zero for x is equal to 0.39894/0.004432 or 90.0 times the probability of measuring $\bar{x} \pm 3s$. Several calculated probabilities for values of x relative to a probability of one for $\bar{x} \pm 3s$ are shown in column C. Column D presents the relative probabilities of making specific measurements when compared to the assigned value of one for the probability of making a measurement of zero for x. Column E contains the relative probabilities based on a probability of one for $\bar{x} \pm 2s$.

There will always be deviations from the Gaussian curve when a finite set of data is used to produce the plot. A plot of the relative probabilities of measuring various values of x will become smoother and will more closely mimic the exponential function as the number of measured data points is increased. An infinite set of data will produce the graph and function known as a Gaussian distribution. Chart 3.5A illustrates a smooth plot of the function generated by calculating the ordinate values of 31 points as opposed to Chart 3.5 which was produced with 13 values. This function shows a high probability of measuring values close to $\bar{x}$ which anticipates that the experimentally measured values of x will cluster about $\bar{x}$.

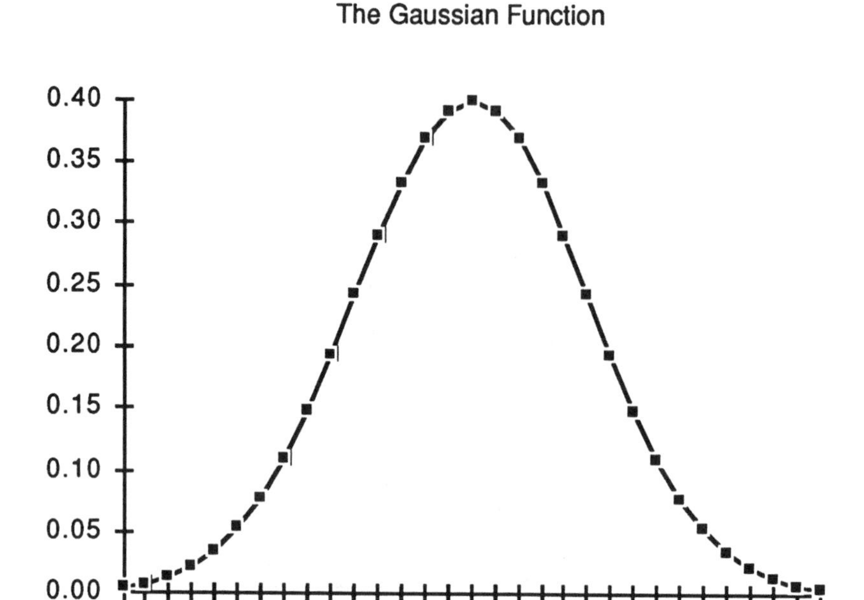

Chart 3.5A The Gaussian Function

The Application of Statistical Analysis

Worksheet 3.7 illustrates the application of statistical analysis to a set of data that contains 13 measurements. The Count(*range*) function is used to show the number of measurements within a given range. The percentage of the measurements lying within that specified range is calculated. The mean or average value and the standard deviation of the measured values are presented in cells B18 and B19. These values establish the range within which 68.26% of the measurements should reside. Column C uses a series of nested IF functions to establish which measured values lie within this range. From these values a numerical count determines the percentage of points that are within one standard deviation of the mean. The result is higher than the theoretically expected value, but you should consider that the data consists of only 13 points (in order for 68.26% of the points to be within this range there would have to be 8.874 points within one standard deviation). Theoretically with an infinite set of data, 95.46% of the measured values should be within a range specified by two standard deviations and 99.74% should be within three standard deviations.

	A	B	C
1	**Statistical Analysis**		
2	Measurement	**Values**	Within a Std Dev
3	1	36.986	
4	2	36.99	
5	3	36.992	36.992
6	4	36.995	36.995
7	5	36.996	36.996
8	6	36.999	36.999
9	7	37	37
10	8	37.001	37.001
11	9	37.004	37.004
12	10	37.005	37.005
13	11	37.006	37.006
14	12	37.009	
15	13	37.014	
16	Count	13	9
17	Percent	100.0%	69.2%
18	Average	37.000	37.000
19	Std Deviaton	0.008	
20	Range	36.992	37.008

Worksheet 3.7 A Practical Application of Statistics

	A	B	C
1	**Statistical**	**Analysis**	
2	Measurement	**Values**	Within a Std Dev
3	1	36.986	=IF(B3<$B$20," ",IF(B3>$C$20," ",B3))
4	=A3+1	36.99	=IF(B4<$B$20," ",IF(B4>$C$20," ",B4))
5	=A4+1	36.992	=IF(B5<$B$20," ",IF(B5>$C$20," ",B5))
6	=A5+1	36.995	=IF(B6<$B$20," ",IF(B6>$C$20," ",B6))
7	=A6+1	36.996	=IF(B7<$B$20," ",IF(B7>$C$20," ",B7))
8	=A7+1	36.999	=IF(B8<$B$20," ",IF(B8>$C$20," ",B8))
9	=A8+1	37	=IF(B9<$B$20," ",IF(B9>$C$20," ",B9))
10	=A9+1	37.001	=IF(B10<$B$20," ",IF(B10>$C$20," ",B10))
11	=A10+1	37.004	=IF(B11<$B$20," ",IF(B11>$C$20," ",B11))
12	=A11+1	37.005	=IF(B12<$B$20," ",IF(B12>$C$20," ",B12))
13	=A12+1	37.006	=IF(B13<$B$20," ",IF(B13>$C$20," ",B13))
14	=A13+1	37.009	=IF(B14<$B$20," ",IF(B14>$C$20," ",B14))
15	=A14+1	37.014	=IF(B15<$B$20," ",IF(B15>$C$20," ",B15))
16	Count	=COUNT(B3:B15)	=COUNT(C5:C13)
17	Percent	1	=C16/B16
18	Average	=AVERAGE(B3:B15)	=AVERAGE(C5:C13)
19	Std Deviaton	=STDEV(B3:B15)	
20	Range	=B18-B19	=B18+B19

Worksheet 3.7F Formulas for Worksheet 3.7

Problems

1. Calculate the atomic weightof tin. The mass numbers, isotopic abundance values in atomic percent, and the relative isotopic mass based on carbon-12 of the ten stable, naturally occurring isotopes of tin are: 112(1.0%) 111.90483, 114(0.7%) 113.902784, 115(0.4%) 114.903348, 116(14.7%) 115.901747, 117(7.7%) 116.902956, 118(24.3%) 117.901609, 119(8.6%) 118.903310, 120(32.4%) 119.902200, 122(4.6%) 121.903440, 124(5.6%) 123.90527.

2. Calculate the atomic weight of oxygen. The mass numbers, isotopic abundance values, and the relative isotopic mass of the three stable, naturally occurring isotopes of O are: 16(99.762%) 15.99491461, 17(0.038%) 16.999131, 18(0.200%) 17.999160.

3. Calculate the atomic weight of sulfur. The mass numbers, isotopic abundance, and the relative isotopic mass of the four stable, naturally occurring isotopes of sulfur are: 32(95.02%) 31.972071, 33(0.75%) 32.971459, 34(4.21%) 33.9678667, 36(0.02%) 35.9670806.

4. Calculate the atomic weight of iron. The mass numbers, isotopic abundance, and the relative isotopic mass of the four stable, naturally occurring isotopes of iron are: 54(5.8%) 53.939613, 56(91.72%) 55.934940, 57(2.2%) 56.935396, 58(0.28%) 57.933278.

5. Calculate the atomic weight of bromine. The mass numbers, isotopic abundance, and the relative isotopic mass of the two bromine isotopes are: 79(50.69%) 78.918336, 81(49.31) 80.91629.

6. Calculate the atomic weight of ruthenium. The mass numbers, isotopic abundance, and the relative isotopic mass of the seven ruthenium isotopes are: 96(5.52%) 95.90760, 98(1.88%) 97.90529, 99(12.7%) 98.905938, 100(12.6%) 99.904218, 101(17.0%) 100.905581, 102(31.6%) 101.904348, 104(18.7%) 103.90542.

7. What is the percent composition by mass of the following sulfur containing compounds? Which compound has the largest percentage of sulfur? Na_2SO_4, $Na_2S_2O_3$, $Na_2S_4O_6$, Na_2SO_3, SO_3, and P_2S_3.

8. What is the percent composition by mass and which compound has the largest percentage of phosphorous? Na_3PO_4, $Na_3P_2O_7$, Na_2HPO_2, P_5S_7, and H_3PO_3.

9. What is the percent composition by mass of the following iodine containing compounds? Which compound has the largest percentage of iodine? $NaIO_4$, $NaIO_7$, NaI, PH_2I, IF_3, and PI_2F.

10. What is the percent composition by mass of the following nitrogen containing compounds? Which compound has the largest percentage of nitrogen? NF_3, N_2O_4, N_2F_2, ClN_3, NH_4F, and NaN_3.

11. The minting of U.S. dimes causes deviations in the mass of the product. Normal wear accounts for the remainder of the deviations in the mass of U.S. dimes. Calculate $\bar{x}$ and s for the measurements of the mass of ten U.S. dimes. Data (in mg): 2235.1, 2250.7, 2251.6, 2254.3, 2256.2, 2247.1, 2257.7, 2277.7, 2235.5, and 2255.2.

12. Precipitation of silver chloride for determining the silver content in a mineral sample produced measurements of 432.5 mg, 433.1 mg, and 431.7 mg when the same mass of mineral is used for each analysis. Calculate the mean and standard deviation of these measurements.

13. In problem 12, the determination of the silver content in a mineral sample, one of the measurements is suspect and requires the analyst to perform three more determinations. Calculate the mean and standard deviation for the six analyses. The original measurements and the new measurements are (in mg): 432.5, 433.1, 431.7, 432.1, 431.6, and 432.3.

14. An ore sample that contains gold was analyzed by five separate and independent determinations. Calculate the mean and standard deviation for the following five measurements: 4.35%, 4.28%, 4.32%, 4.31%, and 4.33%.

15. Density measurements of an alloy resulted in the following measurements (in g/mL): 4.453, 4.424, 4.438, 4.419, and 4.429. Calculate the mean and standard deviation for this data.

16. An analysis for the protein content of a cheddar cheese soup produced the following data: 3.23%, 3.15%, 3.29%, 3.14%, and 3.19%. Calculate the mean and standard deviation for the percentage of protein in this cheddar cheese soup.

17. An analysis for the carbohydrate content of a cheddar cheese soup produced these results: 9.03%, 8.95%, 8.91%, 8.98%, and 9.05%. Calculate the mean and standard deviation for the percentage of carbohydrate in this cheddar cheese soup.

18. An analysis by laser excited fluorescence for the rhenium content of a concentrated ore sample produced the following results (in ppm): 955, 1206, 1301, 1016, 921, 1267 . Calculate the mean and standard deviation in ppm for the content of rhenium in this ore sample.

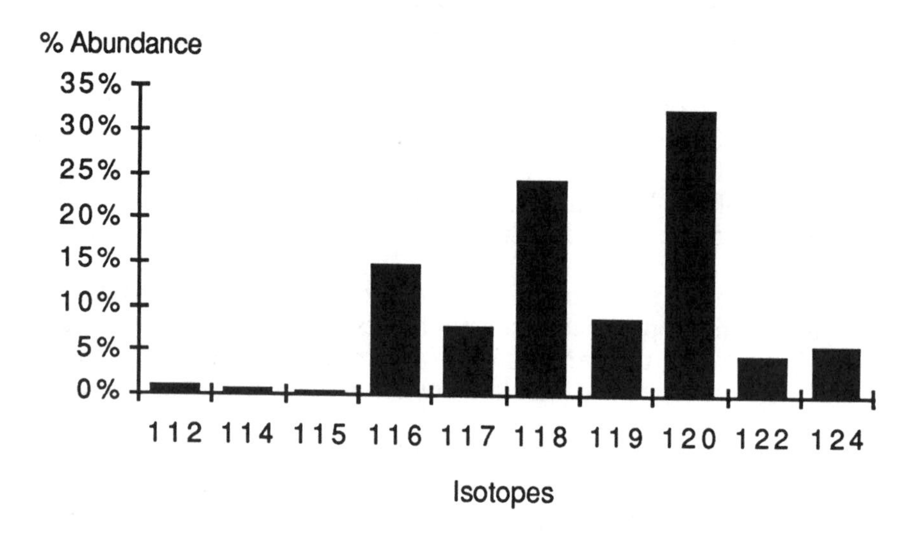

Chart 3.8 Abundance of Sn Isotopes in Problem 1

4

Molecular Weight, the Mole, and Percent Composition

Atoms and molecules are extremely small and have very little mass. At first scientists were only able to talk about their relative mass. Before the turn of the century, they had developed a relative atomic and molecular weight scale and a unit called the mole that let them compare numbers of atoms and molecules even though they did not know the absolute mass or number of the atoms involved. We still use these ideas today because they have been found to be quite serviceable. (BLB Chap. 3)

Molecular Weight

The current atomic weight scale is defined by setting the mass of the isotope carbon-12 exactly equal to 12. All other isotopes are measured experimentally relative to this one. The atomic weights of the isotopes are averaged over all the isotopes of a particular element to get the average atomic weight of that element as it naturally occurs. These relative average atomic weights are the ones we use for most chemical purposes.

Compounds are put on the same relative weight scale by adding together the atomic weights of all the atoms in the formula for the compound. These are called molecular weights. Technically, if no molecule exists, as in ionic compounds, they are called formula weights, but most chemists do not bother with this distinction.

Let's calculate the molecular weight of the compound H_2SO_4. We need to look up the atomic weights of H, S, and O and then count how many atoms are in the formula.

2 H	2 x	1.01	=	2.02
1 S	1 x	32.06	=	32.06
4 O	4 x	16.00	=	64.00
molecular weight				98.08

The units for these atomic weights and the resulting molecular weight were originally unknown and were called atomic mass units, or amu. They did allow for comparison, however. For example, we can say the mass of an oxygen atom is about 16 times as large as the mass of a hydrogen atom. The sulfuric acid molecule used for our example above has a mass of about 98 times that of a H atom.

You must be very careful in counting the number of atoms if parentheses are involved in the formula. For example, $Ca_3(PO_4)_2$ contains three calcium atoms, two phosphorus atoms, and eight oxygen atoms.

Set up a sheet for calculating molecular weight like that shown in Worksheet 4.1. Type labels in the first four rows as shown. In column A starting with cell A5, type the symbols for several elements. Next to each symbol, type the atomic weights for your elements in column B. In cell D5 type =B5*C5. This multiplies the atomic weight of hydrogen by the number of hydrogen atoms you have entered in cell D5. Now select this cell and copy it by selecting Edit, Copy. Next paste this formula into cells D6 through D14 by darkening these cells with the mouse or keyboard and then selecting Edit, Paste.

	A	B	C	D	E
1	**Molecular Weight and Moles**				
2			**Enter**		*Total*
3			**number**		*molecular*
4	*element*	*atomic wt*	**of atoms**	*weight*	*weight*
5	H	1.0079	2	2.0158	98.0734
6	C	12.0111		0	
7	N	14.0067		0	
8	O	15.9994	4	63.9976	
9	Na	22.98977		0	
10	P	30.97376		0	
11	S	32.06	1	32.06	
12	Cl	35.453		0	
13	K	39.0983		0	
14	Ca	40.08		0	

Worksheet 4.1 Spreadsheet for Molecular Weight Calculations

The final step is to type the formula for summing all the weights in column D into cell E5.

=SUM(D5:D14)

Now enter the number 2 in cell C5, the number 4 in cell C8, and the number 1 in cell C11. This corresponds to the formula H_2SO_4 and the molecular weight of this compound should appear in cell E5 as 98.0734. If this works correctly, your sheet is ready to calculate the molecular weight of other compounds. Clear the current compound by darkening cells C5 through C14 (shaded in Worksheet 4.1). Then select Edit, Clear, OK to remove the compound. This should set column D to all zeros along with the molecular weight. Now type in your new numbers corresponding to the number of atoms of each element in your formula.

1. Calculate the molecular or formula weight of each of the following compounds:

a) NaCl
b) H_2O
c) Na_3PO_4
d) C_2H_5OH
e) HNO_3
f) $C_{12}H_{22}O_{11}$

Now let's add some more atomic weights to the sheet. Start adding elements in cell A15 and continue on down the column for each additional one as in Worksheet 4.2. Then copy the formula in cell D14 into cells D15 through D19. Lastly, change the formula in cell E5 to =SUM(D5:D19). You are now ready to work the following problem.

2. Which of the following compounds has the highest molecular weight? Calculate each one to be sure.

a) K_2PtCl_6
b) UF_6
c) $FeBr_4$
d) Al_2Cl_6
e) $Ca_3(PO_4)_2$

More atomic weights can be added or some of the existing ones can be changed to other elements if desired to work other problems. Add elements to Worksheet 4.2 or change elements to work the following problems.

3. Calculate the molecular weights of the following:

a) AuI
b) $[Co(NH_3)_3BrCl_2]$
c) $XeOF_4$
d) $Pb_2[NiCl_6]$

	A	B	C	D	E
1	Molecular Weight and		Moles		
2			Enter		*Total*
3			number		*molecular*
4	*element*	*atomic wt*	of atoms	*weight*	*weight*
5	H	1.0079		0	162.206
6	C	12.0111		0	
7	N	14.0067		0	
8	O	15.9994		0	
9	Na	22.98977		0	
10	P	30.97376		0	
11	S	32.06		0	
12	Cl	35.453	3	106.359	
13	K	39.0983		0	
14	Ca	40.08		0	
15	Fe	55.847	1	55.847	
16	U	238.029		0	
17	Al	26.98154		0	
18	Pt	195.09		0	
19	F	18.998403		0	

Worksheet 4.2 Adding More Elements to the Sheet

Percent Composition

The typical elemental analysis results reported for a new compound are given in percent composition by weight for each of the elements contained in the compound. For example water, H_2O, contains 11.2% H and 88.8% O. Of course, the formulas indicate how much of each element is contained by number of atoms. Water contains 2 H and 1 O. These two ways of indicating the contents of a compound are different because the atoms of one element have a different mass than atoms of the other elements. The amount expressed one way can be converted to the other, however.

It is very useful when dealing with formulas and balanced chemical equations to be able to compare the number of atoms and molecules or formula units. The unit called the mole (mol) was developed for this. It originally was defined as the amount of material in grams that was numerically equal to the atomic or molecular weight. Thus, one could use the units g/mol on molecular and atomic weights. Equal numbers of moles then had an equal number of atoms or molecules, but the exact number in a mole was not known. It is still a useful number for many purposes, such as the following.

To convert from the formula to percent composition, assume you have one mole of material. The total weight of the material now just equals the molecular weight, which you can deduce from the formula. The weight of each element will just equal the weight of the number of moles of that element in one mole of the compound and will equal the number of atoms of that element in the formula times the atomic weight of that element. The percent of element, A, in the compound will then be:

$$\%A = n \times AW(A) \times 100 / MW(\text{compound})$$

where n is the subscript on element A in the formula, AW is the atomic weight of element A, and MW is the molecular or formula weight of the compound.

For example, consider percent hydrogen and oxygen in water.

$$\%H = 2 \times 1.01 \times 100 / 18.02 = 11.2\%$$
$$\%O = 1 \times 16.00 \times 100 / 18.02 = 88.8\%$$

Let's add a column to our spreadsheet for calculating molecular weight so we can also calculate percent composition. Worksheet 4.3 shows this addition in shading. Place the formula =100*$D5/$E$5 into cell F5. Copy this cell into cells F6 through F19. Notice the absolute cell reference for E5. This is to ensure division is always done using the molecular weight of the compound. Now use this modified sheet to work the following problems. Ignore the divide by zero errors when you clear the number of atoms out of column C. This is due to the molecular weight being zero until you enter some new atoms.

4. Calculate the percent of the elements indicated for each compound:

a) Percent Cl in K_2PtCl_6
b) Percent U in UF_6
c) Percent Br in $FeBr_4$
d) Percent Al in Al_2Cl_6
e) Percent O in $Ca_3(PO_4)_2$
in H_2O
in C_2H_5OH
in HNO_3
in $C_{12}H_{22}O_{11}$

	A	B	C	D	E	F
1	**Molecular Weight and Moles**					
2			**Enter**		*Total*	*percent*
3			**number**		*molecular*	*composi-*
4	*element*	*atomic wt*	**of atoms**	*weight*	*weight*	*tion*
5	H	1.0079	2	2.0158	18.0152	11.18944
6	C	12.0111		0		0
7	N	14.0067		0		0
8	O	15.9994	1	15.9994		88.81056
9	Na	22.98977		0		0
10	P	30.97376		0		0
11	S	32.06		0		0
12	Cl	35.453		0		0
13	K	39.0983		0		0
14	Ca	40.08		0		0
15	Fe	55.847		0		0
16	U	238.029		0		0
17	Al	26.98154		0		0
18	Pt	195.09		0		0
19	F	18.998403		0		0

Worksheet 4.3 Calculating Percent Composition

Gram, Mole, Molecules Conversions

We now know that one mole equals 6.02×10^{23} atoms or molecules (Avogadro's number, N_o). The mole is very much like counting in dozens except the conversion factor is N_o instead of 12.

It is possible to convert the amount of a material between grams, moles, and molecules (or atoms) using the molecular and atomic weights and Avogadro's number. The diagram below shows the conversion factors for these conversions.

grams <——————> moles <—————> molecules

molecular weight (g/mol) — 6.02×10^{23} (molecules/mol)

Let's look at an example of these conversions. How many moles and molecules are in 100.0 grams of H_2O? We will need the molecular weight of water.

2 H = 2 x 1.01 = 2.02
1 O = 1 x 16.00 = 16.00
18.02 g/mol

Now use this conversion factor to determine the number of moles of water.

100.0 g / 18.02 g/mol = 5.55 mol

Notice we had to divide by the molecular weight for the units to come out correctly, just like in any other unit conversions you have done. Of course, you will multiply by the molecular weight to convert moles to grams.

Next use Avogadro's number to determine the number of water molecules.

5.55 mol x 6.02×10^{23} molecules/mol = 3.34×10^{24} molecules

Here we had to multiply by the conversion factor to get the correct units. We of course will divide by N_o when converting from molecules to moles.

Make the following additions to your sheet so you can input the amount of a compound in one unit and calculate the amount in another. See Worksheet 4.4 for an example.

Enter 'Grams, moles, molecules' into cell G2.
Enter 'Amount of compound:' into cell G3.
Enter '<-- enter "moles"', into cell H4.
Enter '"grams" or "molecules"' into cell G5.
Enter '<--enter numeric amount' into cell H6.
Enter 'value' into cell I7.

The last four cells we will enter involve very complicated IF statements so that different calculations and labels can be used depending on what units you use to enter the amount of material (see shaded cells in Worksheet 4.4). So be <u>very careful</u> in typing them, paying special attention to all punctuation.

In cell G8:

```
=IF(OR(G4="moles",G4="grams",G4="molecules"),(IF(G4="grams",G6/
    E5,(IF(G4="moles",G6*E5,G6/6.02E+23)))),"error")
```

	A	B	C	D	E
1	**Molecular Weight and Moles**				
2			**Enter**		*Total*
3			**number**		*molecular*
4	*element*	*atomic wt*	**of atoms**	*weight*	*weight*
5	H	1.0079	2	2.0158	18.0152
6	C	12.0111		0	
7	N	14.0067		0	
8	O	15.9994	1	15.9994	
9	Na	22.98977		0	
10	P	30.97376		0	
11	S	32.06		0	
12	Cl	35.453		0	
13	K	39.0983		0	
14	Ca	40.08		0	
15	Fe	55.847		0	
16	U	238.029		0	
17	Al	26.98154		0	
18	Pt	195.09		0	
19	F	18.998403		0	

	F	G	H	I
1				
2	*percent*	***Grams, moles, molecules***		
3	*composi-*	Amount of compound:		
4	*tion*	grams	**<-- enter "moles",**	
5	11.18944		**"grams" or "molecules"**	
6	0	100	**<-- enter numeric**	
7	0			**value**
8	88.81056	5.5508682	**moles**	
9	0	3.342E+24	**molecules**	
10	0			
11	0			
12	0			
13	0			
14	0			
15	0			
16	0			
17	0			
18	0			
19	0			

Worksheet 4.4 Conversion Among Grams, Moles, and Molecules

In cell H8:

```
=IF(OR(G4="moles",G4="grams",G4="molecules"),IF(G4="grams","
    moles",IF(G4="molecules","moles","grams")),"error")
```

In cell G9:

```
=IF(OR(G4="moles",G4="grams",G4="molecules"),(IF(G4="grams",G
  8*6.02E+23,(IF(G4="moles",G6*6.03E+23,G8*E5)))),"error")
```

In cell H9:

```
=IF(OR(G4="moles",G4="grams",G4="molecules"),IF(G4="grams","
  molecules",IF(G4="molecules","grams,"molecules")),"error")
```

Try entering 100 grams of water and see if you get the results in the previous example. This will let you know if you have typed the above formulas correctly. Then work the following problems.

5. Calculate the number of moles and molecules in 354.8 grams of K_2PtCl_6.

6. Calculate the number of grams and molecules in 2.453 moles of H_2O.

7. Calculate the number of grams and moles in 9.12×10^{12} molecules of Al_2Cl_6.

8. Calculate the number of moles and atoms in 529.0 grams of Fe. (Treat this as a one-atom molecule.)

9. What is the mass in grams of one C_2H_5OH molecule?

10. What is the mass in grams of one U atom?

11. What is the mass of 25 formula units of NaCl?

5

Empirical and Molecular Formulas

The ratio of atoms of the different elements in a new compound (the empirical formula) can be determined from experimental percent composition data. The actual number of atoms in a molecule (molecular formula) can also be determined for molecular compounds if the molecular weight has been experimentally determined. (BLB Chap. 3)

The typical elemental analysis results reported for a new compound are given in percent composition by weight for each of the elements contained in the compound. For example water, H_2O, contains 11.2% H and 88.8% O. Of course, the formula indicates how much of each element is contained in terms of the number of atoms. Water contains two H atoms and one O atom. These two ways of indicating the contents of a compound are different because the atoms of each element have a different mass than atoms of the other elements. The atom ratio was converted to percent composition in the last chapter.

To find the formula from the percent composition, assume you have 100 g of material. This means the number of grams of each element just equals the percentage. The number of grams are then converted to moles by dividing each mass by the atomic weight of that element. The mole ratios are then determined to get the relative number of atoms of each type. This comparison can be done by dividing all of the mole values by the smallest of them.

If the results are not all close to integer values, multiply them all by the number required to convert them all to integers. For example let's look at a phosphorus oxide that contains 43.7% P and 56.3% O. Assume we have 100 grams of the compound. Then grams equal the percent.

$$\text{grams P} = 100 \times 43.7/100 = 43.7$$
$$\text{grams O} = 100 \times 56.3/100 = 56.3$$

Moles equal the grams divided by the atomic weight.

moles P = 43.7/31.0 = 1.41
moles O = 56.3/16.0 = 3.52

Compare moles by dividing each value by the smallest, 1.41 in this case.

relative mol P = 1.41/1.41 = 1.00
relative mol O = 3.52/1.41 = 2.50

Since 2.5 is not close to integer, multiply all relative mole values by 2. This makes the 2.5 become 5 and the other value still remains an integer, 2.

integer mol P = 1.00 x 2 = 2.00
integer mol O = 2.50 x 2 = 5.00

So the empirical formula for this oxide of phosphorus is P_2O_5. Note that if the experimental data is not very accurate, the value will not come out exactly integer. In cases where the data is extremely inaccurate, it may be difficult to decide what the formula should be.

This empirical formula may not be the correct formula for the molecule. The real molecule may be some multiple of this empirical formula. For example, the empirical formula for benzene is CH, while the molecular formula is C_6H_6, a sixfold multiple of the empirical formula. To determine the molecular formula from the empirical formula, one more piece of data is needed: the molecular weight. We will see later in the chapter how this is used.

Worksheet 5.1 shows the start of a spreadsheet for determining empirical formula from percent composition. Put your headings in the cells shown and a list of elements in cells A5 through A14. Cells B5 through B14 contain the atomic weights of these elements. The shaded area, cells C5 through C14, are where you will enter your percent composition data. Worksheet 5.1 has the data for our previous example entered.

Cells D5 through D14 are where moles are calculated by diving the grams (equal to the percentages) by the atomic weight. To set this calculation up enter the following formula into cell D5:

```
=IF(ISBLANK(C5),"",C5/B5)
```

The IF function checks to see whether the cell containing the percentage for that element is blank. If it is a blank then the calculation is not done and the empty string "" is displayed; otherwise the number of moles is displayed. Copy this formula from cell D6 through cell D14.

	A	B	C	D
1	**Empirical Formula**			
2			**percent**	
3			**compos-**	
4	*element*	*atomic wt*	**ition (%)**	*moles*
5	H	1.0079		
6	C	12.0111		
7	N	14.0067		
8	O	15.9994	56.3	3.518882
9	Na	22.98977		
10	P	30.97376	43.7	1.4108717
11	S	32.06		
12	Cl	35.453		
13	K	39.0983		
14	Ca	40.08		

Worksheet 5.1 Start of Spreadsheet for Determining Empirical Formula

	A	B	C	D	E
1	**Empirical Formula**				
2			**percent**		
3			**compos-**		*mole*
4	*element*	*atomic wt*	**ition (%)**	*moles*	*ratio*
5	H	1.0079			
6	C	12.0111			
7	N	14.0067			
8	O	15.9994	56.3	3.518882	2.4941191
9	Na	22.98977			
10	P	30.97376	43.7	1.4108717	1
11	S	32.06			
12	Cl	35.453			
13	K	39.0983			
14	Ca	40.08			

Worksheet 5.2 Calculating the Relative Number of Moles

Worksheet 5.2 shows another column added that calculates the relative number of moles of each element compared to the smallest one. Type the following into cell E5:

=IF(D5="","",D5/MIN($D$5:$D$14))

Again a check is made for a number in the adjacent cell and no calculation is done if a blank is found. The MIN function finds the smallest mole value from the list in column D. Copy cell E5 into cells E6 through E14.

If the mole ratio does not come out close to an integer for all elements in the compound the values are multiplied by a factor that you enter into the shaded cell F3 shown in Worksheet 5.3. All values in column E are multiplied by this value and entered into column F. Type the following into cell F5:

=IF(E5="","",E5*$F$3)

Again a check is made for blank cells and the results are put in column F.

In column G the results in F are checked to see if all the numbers are within 0.1 of being an integer. If they are they are rounded off and displayed. If they are not, the message "no integer" is placed in the cell. Put the following formula in cell G5:

=IF(F5="","",IF(ABS(F5-INT(F5+0.5))<0.1,INT(F5+0.5),"no integer"))

Copy this cell into cells G6 through G14. If your sheet now matches Worksheet 5.3, you are ready to work the following problems. To work problems involving other elements you can replace some of the elements and atomic weights in columns A and B. Or you can add onto columns A and B. Just be sure you also extend the other columns by copying the formulas on down each column.

	A	B	C	D	E	F	G
1	Empirical Formula					multiplying	
2			percent			factor	
3			compos-		*mole*	2	final
4	*element*	*atomic wt*	ition (%)	*moles*	*ratio*	new ratio	ratio
5	H	1.0079					
6	C	12.0111					
7	N	14.0067					
8	O	15.9994	56.3	3.518882	2.4941191	4.9882382	5
9	Na	22.98977					
10	P	30.97376	43.7	1.4108717	1	2	2
11	S	32.06					
12	Cl	35.453					
13	K	39.0983					
14	Ca	40.08					

Worksheet 5.3. Finished Spreadsheet for Determining Empirical Formula from Percent Composition

1. What is the empirical formula for each of the following compounds?

a) 20.1% Ca and 79.9% Br
b) 37.5% C, 12.6% H, and 49.9% O
c) 16.1% K, 40.1% Pt, and 43.8% Cl
d) 47.1% C, 2.48% H, 26.9% Mn, and 23.5% O

Molecular Formulas

The molecular formula for your compound can be determined from your empirical formula if you know the molecular weight of your compound. First calculate the weight of your empirical formula by adding up the atomic weights of all the atoms in the formula. Compare this to the molecular weight. If they are approximately the same, then the empirical formula is also the molecular formula. This situation is very common, especially for small molecules.

If the values differ, the molecule is actually some integer number multiple of the empirical formula. Divide the molecular weight by the empirical formula weight to see what this multiple is. Multiply all the subscripts in the empirical formula by this multiple to obtain the molecular formula. For example, if your molecular formula was CH (weight = 13.0 g/mol) and the molecular weight was 26.0 g/mol, multiply the coefficients of the molecular formula by 26.0/13.0 = 2 to obtain the molecular formula of C_2H_2.

Worksheet 5.4 shows the previous spreadsheet extended to carry out this calculation. Type the following formula into cell H5:

```
=IF(G5="","",G5*B5)
```

This multiplies the number of moles for each element by the atomic weight so in the next column the weight of the empirical formula can be determined. Copy cell H5 into cells H6 through H14. The molecular weight is entered in cell I2. Then in column I, the number of moles in the actual molecule are determined. Enter the following formula into cell I5:

```
=IF($G5="","",INT($G5*$I$2/SUM($H$5:$H$14)+0.5))
```

This formula calculates the molecular formula subscripts and converts them to integer values. The SUM function is determining the total empirical formula weight from column H. Copy this formula into cells I6 through I14.

Try this addition to your sheet using 7.76% H and 92.24% C and a molecular weight of 26.02 g/mol. This should lead to the molecular formula in the preceding example. If you modify this sheet for other elements by extending all the columns, be sure you also change the range of cells in the SUM function in cell I5. Then copy this new formula the rest of the way down column I. Now work the following problems.

	A	B	C	D	E
1	Molecular Formula				
2			percent		
3			compos-		mole
4	element	atomic wt	ition (%)	moles	ratio
5	H	1.0079	7.76	7.6991765	1.002554
6	C	12.0111	92.24	7.6795631	1
7	N	14.0067			
8	O	15.9994			
9	Na	22.98977			
10	P	30.97376			
11	S	32.06			
12	Cl	35.453			
13	K	39.0983			
14	Ca	40.08			

	F	G	H	I
1	multiplying	Molecular Weight		
2	factor		--------->	26.02
3	1	Empirical	Empirical	Molecular
4	new ratio	Formula	Weight	Formula
5	1.002554	1	1.0079	2
6	1	1	12.0111	2
7				
8				
9				
10				
11				
12				
13				
14				

Worksheet 5.4 Spread sheet for Determining Molecular Formula from Percent Composition and Molecular Weight

2. What is the molecular formula for each of the following compounds?

a) 85.6% C and 14.4% H; molecular weight 28.04
b) 85.6% C and 14.4% H; molecular weight 85.2
c) 20.2% Al and 79.8% Cl; molecular weight 267.0
d) 40.0% C, 6.73% H, and 53.3% O; molecular weight 60.06

6

Stoichiometry and Titration

Stoichiometry involves using the mole relationships in balanced chemical equations to calculate amounts of some of the materials in a reaction from given amounts of other materials. These calculations have a broad range of applications from determining yield of products, predicting the amount of pollutants, relating quantities in equilibrium calculations, doing titration calculations, and gravimetric analysis, to name a few. (BLB Chap. 3 and 4)

The details of many problems can be quite complicated, but central to every one of them are the same few straightforward steps. At the heart of the whole procedure is the balanced chemical equation. Balancing an equation reflects the fact that during chemical changes, the various atoms themselves are left unchanged. They are only rearranged with small, but important, changes in the electron distribution around them. Thus, the total mass remains unchanged during the process. The coefficients in the balanced equation reflect this and can be interpreted as the relative number of different formula units or molecules involved in a reaction. These numbers can also be, and usually are, interpreted as the number of moles of each material for most stoichiometry uses.

At first sight there appear to be four distinct types of relationships between the materials involved in a chemical reaction.

1. How much product B can be produced from a given amount of reactant A?

$$aA + ... \rightarrow bB + ...$$

where a and b are the coefficients in the balanced equation.

2. How much reactant B is needed to produce a given amount of product A?

$$bB + ... \rightarrow aA + ...$$

3. How much reactant B is needed to react with a given amount of reactant A?

$$aA + bB + \ldots \rightarrow \ldots$$

4. How much product B is produced when a given amount of product A is produced?

$$\ldots \rightarrow aA + bB + \ldots$$

While these different questions appear different and in many practical respects are asked for a variety of different reasons, the solution to all of these follows one simple pattern. The distinction between reactant and product is not necessary for solving these problems, although it may be crucial in interpreting the results and determining how they are used.

Let's consider the solution to all of these problems by looking at the following four steps:

1. Having a properly balanced chemical equation is essential since the coefficients give us all of the relationships between the amounts of the various materials. Actually only the coefficients of the materials given and sought need to be known, but typically the others are determined in the balancing process.

2. To make use of these relationships, the amount of material given must be converted into moles. This might be a simple gram to mole conversion, a molarity and volume to mole conversion, or something even more complicated, such as calculating the results for another stoichiometry problem. This can be a simple step or the most difficult part of the problem.

3. The coefficients in the balanced equation relate the amounts of the materials to each other regardless of which side of the reaction they are on, so the reactant-product distinction is not needed. The amount of the sought after material can always be calculated from a ratio of the coefficients of the two materials involved in the problem. Note that in the above general reactions, the given material was always compound A and the sought after material was compound B. So the same ratio can be used to work all four types.

$$(\text{moles A}) \times b/a = (\text{moles B})$$

4. The last step involves converting the moles of the sought after material, B, to whatever units are required by the particular problem.

Like step 2 above, this may be straightforward or very complicated, depending on the individual problem.

These identical four steps can be used to solve all of the apparently different four types of problems above. Let's look at an example. How many grams of ammonia, NH_3, can be produced starting with 33.0 g of H_2, using the reaction:

$$H_2 + N_2 \rightarrow NH_3$$

1. Balance the equation if it is not already balanced.

$$3\ H_2 + N_2 \rightarrow 2\ NH_3$$

2. Convert the amount of the given material, H_2, to moles.

$$33.0\ g\ /\ 2.02\ g/mol = 16.3\ mol\ H_2$$

3. Use the coefficients from the balanced equation to calculate moles of the sought after material, NH_3.

$$16.3\ mol \times (2/3) = 10.9\ mol\ NH_3$$

Be very careful to write the ratio the correct way: sought after material over given material.

4. Convert amount of ammonia to grams.

$$10.9\ mol \times 17.0\ g/mol = 185\ g\ NH_3$$

Let's set up a sheet to do these types of calculations as shown in Worksheet 6.1. We will set it up so you can have as many as a total of eight reactants and products. The formulas can be put in row 5 with the coefficients in the balanced equation just below each formula in row 6. Molecular weights are entered below in row 7. The amount of the given material in grams will be entered in cell B8 under the column headed "given." This can be either a reactant or a product. The rest of the reactants and products can be listed in any order in columns D through J. Moles of the given material are calculated in cell B9 (=$B$8/$B$7).

The grams and moles of the other materials are calculated in rows D through J. Type the following formula into cell D8:

```
=D$6*$B$9*D$7/$B$6
```

Be careful where you place the $'s. Next, into cell D9, type the formula:

=IF(D$7>0,D$8/D$7,0)

The IF prevents us from trying to divide by zero if the molecular weight has not been entered in each column. Now copy these two cells (D8 and D9) into cells E8 through J9. Data can then be entered into the shaded cells to calculate the amount of any or all of the other reactants and/or products in your balanced reaction. To clear any of these cells for a new calculation, use the Edit, Clear, OK sequence after selecting these cells with the mouse.

	A	B	C	D	E	F	G
1		Stoichiometry					
2							
3	**Given Grams of one material**				**Amounts of other mat**		
4		given		calc	calc	calc	calc
5	formula -------->	H2		N2	NH3		
6	coefficient ---->	3		1	2		
7	Molec Wgt --->	2.02		28	17		
8	**grams ------>**	33		152.5	185.15	0	0
9	moles	16.34		5.446	10.891	0	0

Worksheet 6.1 Calculating the Amount of Material Involved in a Chemical Reaction Given the Grams of One Material

Compare your sheet's results with the previous example to see if it is set up correctly. Next, work the following three problems using your sheet.

1. How many grams of O_2 are needed to produce 23.6 grams of H_2O using the reaction

$$C_3H_8 + 5\ O_2 \rightarrow 3\ CO_2 + 4\ H_2O$$

2. How many grams of CO are needed to react with 94.6 grams of Fe_2O_3 using the reaction

$$Fe_2O_3 + 3\ CO \rightarrow 2\ Fe + 3\ CO_2$$

3. How many grams of H_2O are produced when 73.2 grams of Cu are produced using the reaction

$$3\ Cu_2O + CH_4 \rightarrow CO + 2\ H_2O + 6\ Cu$$

Limiting Reagent Problems

In some stoichiometry problems, the amounts of two reactants are given and the amount of product that can be produced from this mixture is requested. In general, the two reactants will not be in the same mole ratio as in the balanced chemical reaction so one of the materials will be used up, leaving some of the other reactant left over. The material used up will thus limit how much product can be formed and is called the "limiting reagent."

To calculate the correct amount of product formed, the amount of this limiting reagent must be used in the calculation, so its identity must first be determined. This can be done by comparing the mole ratio of the two reactants in the balanced equation with the actual mole ratio based on the amounts given in the problem.

Another approach that is especially easy, using the above spreadsheet, is to do two separate calculations for the amount of product that could be formed, assuming that each of the two given reactants was completely used up. Two different answers will be obtained, with the smaller of the two being the correct one. The smaller answer is correct because it will involve the material that will be used up first in the mixture. Use this approach to solve the following problems by calculating the amount of product formed from each of the two amounts of reactants given, and then throw away the larger result.

4. How many grams of AgCl are produced if a mixture of 47.6 g of $AgNO_3$ and 26.4 g of NaCl are reacted together?

$$AgNO_3 + NaCl \rightarrow NaNO_3 + AgCl$$

5. How many grams of H_2 can be produced from 5.00 g of HBr and 7.00 g of Zn using the reaction

$$2\ HBr + Zn \rightarrow ZnBr_2 + H_2$$

Titration Problems

Titrations make use of the stoichiometry of various reactions to analyze samples for the amount of different compounds they contain. Experimentally, a solution of one material is slowly added to a sample with which it will react. The point at which the sample is just used up in the reaction is detected by various means, such as chemical indicators or a pH meter. The method of detecting this "end-point" varies with the reaction being used. In general, we then need to calculate the amount of one material used up when a certain amount of the other material has reacted.

Titrations typically involve the concentrations of one or two solutions whose volumes are measured. Thus, the calculations involve amounts in volume and molarity, a concentration unit. The results of the calculation are also often requested in terms of molarity. Molarity is just the number of moles of a material dissolved in a liter of solution, or:

$$M = n / V(\mathrm{L})$$

where M is the molarity, n is the number of moles, and V is the volume in liters. Our original sheet must be modified to accept data and calculate results in terms of volume and molarity.

Let's set up a sheet that will solve a couple of types of titration problems. One type has the sample amount given in grams and the other type has the sample amount given in terms of volume. The amount of titrant (material used to titrate the sample) is usually specified in terms of volume. The unknown quantity may be the molarity of either the sample or titrant, depending on the particular problem.

Refer to Worksheet 6.2 as you set up a sheet as follows. The coefficient for the titrant goes in cell B6 and for the sample in D6. The two volumes go in cells B7 and D7. The molarity of the titrant is input in cell B8. Moles are calculated in cell B9 (=B7*B8/1000). The sample molarity is calculated in cell D8 [=B7*B8*D6/(B6*D7)]. Moles are calculated in D9 (=D7*D8/1000).

If the sample amount is given in grams and the titrant amount in volume, as happens when you are standardizing a solution, the right side of Worksheet 6.2 can be used. Coefficients go in F6 and H6. The molecular weight of the sample goes in F7 and the volume of titrant in H7. The number of grams of sample is entered in F8. Moles of sample are calculated in F9 (=F8/F7). The molarity of the titrant is calculated in H8 [=H6*F9*1000/(F6*H7)]. Moles of titrant, if needed, are in H9 (=H7*H8/1000).

	A	B	C	D
1		**Titrations**		
2				
3	**Molarity from volume and molarity**			
4		titrant		sample
5	formula ---->	NaOH		H2SO4
6	coefficient >	2		1
7	Volume(mL) ->	19.75		25
8	molarity --->	0.0881		0.0348
9	moles	0.00174		0.00087

	E	F	G	H	I	J
1						
2						
3		**Molarity from grams**				
4		sample		titrant		
5		KHP		NaOH		
6		1		1		
7	MW-->	204.22		22.57	<-Volume(mL)	
8	grams-->	0.325		0.07051	<-molarity	
9		0.001591		0.001591		

Worksheet 6.2 Sheet for Working Titration Problems

Work the following two problems and compare your answers with Worksheet 6.2 to see if your sheet is correct. Then work the remainder of the problems.

6. What is the molarity of a sulfuric acid solution if 25.00 mL of it is titrated with 19.75 mL of .0881 M NaOH?

$$H_2SO_4 + 2\ NaOH \rightarrow Na_2SO_4 + 2\ H_2O$$

7. If 0.325 g of potassium hydrogen phthalate (KHP) were titrated with 22.57 mL of NaOH, what was the molarity of the NaOH if the molecular weight of KHP is 204.22 g/mol?

$$KHP + NaOH \rightarrow NaKP + H_2O$$

8. What is the molarity of an HCl solution if 20.0 mL are titrated with 23.4 mL of 0.165 M NaOH?

$$HCl + NaOH \rightarrow NaCl + H_2O$$

9. What is the molarity of an I_2 solution if 50 mL of it is titrated with 26.4 mL of a 0.0227 M solution of $Na_2S_2O_3$?

$$S_2O_3^{=} + 4\ I_2 + 10\ OH^- \rightarrow 2\ SO_4^{=} + 8\ I^- + 5\ H_2O$$

10. What is the molarity of a $KMnO_4$ solution if 36.7 mL of this solution is used to titrate 0.988 g of $Na_2C_2O_4$?

$$5\ Na_2C_2O_4 + 2\ KMnO_4 + 8\ H_2SO_4 \rightarrow 10\ CO_2 + 2\ MnSO_4 + 8\ H_2O + 5\ Na_2SO_4 + K_2SO_4$$

7

Calorimeter Measurements and Standard Enthalpy of Reaction

All chemical reactions involve heat being released or absorbed. This heat can be measured in a calorimeter. This is an insulated container in which the reaction is run. The temperature in the calorimeter will change during the course of the reaction due to heat being evolved or used. This temperature change can be measured, and if the heat capacity of the calorimeter and its contents is known, the heat can be calculated. (BLB Chap. 5)

The following equation is used to calculate heat from temperature change and the heat capacity of the calorimeter.

$$q = C \times (T_f - T_i) = C \times \Delta T$$

where q is the heat, C is the heat capacity, T_f is the final temperature, and T_i is the initial temperature. The heat capacity of the calorimeter can be determined by electrical heating or by running a reaction whose heat is known.

If the temperature change is positive (increase in temperature), the heat has a positive value. This is the heat entering the calorimeter. This heat left the reactants to heat up the calorimeter, so with respect to the reaction, the heat is negative. In this case heat is produced by the reaction and the reaction is called *exothermic.*

If the temperature change is negative, the calorimeter is giving up heat to the reaction. The reactants are using heat and the heat with respect to the reaction is positive. This type of reaction is called *endothermic.*

The heat measured with the volume of the reactants kept constant is called the change in energy (ΔE).

$$q_v = (E_f - E_i) = \Delta E$$

Heat measured with the pressure kept constant is called the change in enthalpy (ΔH). Many chemical reactions are measured under these constant pressure conditions, so we will be interested mainly in the change in enthalpy.

$$q_p = (H_f - H_i) = \Delta H$$

Sample Problem: 450 grams of a compound whose molecular weight equals 321 g/mol is reacted in a calorimeter whose heat capacity was previously determined to be 897 kJ/K. If the temperature change during the reaction is +5.86 K, how much heat did the calorimeter gain or lose? What is the total heat released or gained by the reactants? What is the enthalpy change per mole of the reactant?

Solution: The heat involved is

$$q = C \times \Delta T = 897 \times (+5.86) = 5256 \text{ kJ}$$

Since the value is positive, this is the heat *gained* by the calorimeter and its contents. This heat is *released* by the reactants, so the heat with respect to the reactants is -5256 kJ, making this an exothermic reaction. The number of moles of reactant is

$$450 \text{ g} / 321 \text{ g/mol} = 1.40 \text{ mol}$$

The enthalpy change per mole of reactant is then

$$q(\text{reaction}) = -q(\text{calorimeter}) = -5256 \text{ kJ}$$
$$\Delta H = -5256 \text{ kJ} / 1.40 \text{ mol} = -3750 \text{ kJ/mol}$$

Set up a sheet like Worksheet 7.1 to solve the previous sample problems. Worksheet 7.1F shows the formulas. Then use this sheet to solve the following problems. Be careful to include a minus sign on temperature change if the temperature decreases.

	A	B	C
1	Calorimetry Calculations		
2			
3	Input:	mass (grams)	450
4		heat cap (kJ/mol)	897
5		Temp change (K)	5.86
6		Molec wight (g/mol)	321
7	Results:	heat (kJ) calorimeter	5256.42
8		heat (kJ) reaction	-5256.42
9	Enthalpy of reaction (kJ/mol)		-3749.5796

Worksheet 7.1 Sheet for Calorimetry Calculations

	A	B	C
1	Calorimetry Calculations		
2			
3	Input:	mass (grams)	450
4		heat cap (kJ/mol)	897
5		Temp change (K)	5.86
6		Molec wight (g/mol)	321
7	Results:	heat (kJ) calorimeter	=C4*C5
8		heat (kJ) reaction	=-C7
9	Enthalpy of reaction (kJ/mol)		=C8*C6/C3

Worksheet 7.1F Sheet Showing Formulas for Calorimetry Calculations in Worksheet 7.1

1. An excess of NaOH is reacted with 25.0 grams of H_2SO_4 in a calorimeter whose heat capacity, including the water in which the reactions is carried out, was previously determined to be 6.46 kJ/K.

$$H_2SO_4 + 2\ NaOH \rightarrow Na_2SO_4 + 2\ H_2O$$

The temperature change is +4.8 K during the reaction. How much heat is evolved during this reaction? What is the ΔH per mole of H_2SO_4 for this neutralization reaction?

2. When $K_2Cr_2O_7$ is dissolved in water, the temperature of the water decreases.

$$K_2Cr_2O_7 \rightarrow 2\ K^+\ (aq) + Cr_2O_7{}^{-2}\ (aq)$$

If this process is carried out in a calorimeter whose heat capacity is 1.16 kJ/K using 10.0 g of $K_2Cr_2O_7$, the temperature change is -2.14 K.

How much heat is involved in this process? What is the heat of solution (enthalpy change per mole) of potassium dichromate?

3. Ethane is burned completely according to the following reaction:

$$C_2H_6 + 7/2\ O_2 \rightarrow 2\ CO_2 + 3\ H_2O$$

What is the heat for this reaction if the temperature change is +9.41 K when 43.7 grams of ethane are burned in a calorimeter whose heat capacity is 242 kJ/K? What is the heat of combustion (enthalpy change per mole) of ethane?

If the data is presented differently in problems (such as giving the initial and final temperatures instead of the temperature change) the sheet needs some modifications. See if you can modify your sheet or design a new one to work the following problems:

4. The combustion of 10.0 g of coke raises the temperature of the calorimeter (heat capacity = 8.37 kJ/K) from 10°C to 47°C. Calculate the heat value of coke in kJ per gram.

$$C_{(s)} + O_{2(g)} \rightarrow CO_{2(g)}$$

5. Carbon reacting with 2.7 mol of CaO produces a temperature change of -7.3 degrees in a calorimeter whose heat capacity is 172 kJ/K. What is the heat used per mole of CaO for this reaction?

$$CaO_{(s)} + 3\ C_{(s)} \rightarrow CaC_{2(s)} + CO_{(g)}$$

6. Fe_2O_3 is reacted with 4.00 g of Al in a calorimeter.

$$2\ Al + Fe_2O_3 \rightarrow 2\ Fe + Al_2O_3$$

The heat for this reaction is 425 kJ/mol Al. What is the heat capacity in kJ/K of the calorimeter if the temperature change for this reaction is 2.37 degrees?

Standard Enthalpy from Heats of Formation

The heat or enthalpy of many different reactions has been measured and tabulated. If all reactions were tabulated this way, the list would be very long. Also, it is impossible to measure the heat of some reactions accurately. Hess' Law of Heat Summation solves these two problems. This law says that the enthalpy change of a reaction is the same no matter what path is used to go between the reactants and products.

Imagine carrying out a reaction by first taking all the reactant compounds and breaking them apart into the elements from which they are made. Then in a second step reassemble these elements into the products. The enthalpy involved is exactly the same as going directly from the reactants to the products using whatever mechanism the real reaction actually uses. If you knew the enthalpy involved in breaking each compound into elements and forming new compounds from the elements, you could calculate the enthalpy of any reaction. This information is available in tables of standard enthalpies of formation. See Table 7.1 for some selected values.

The *standard enthalpy of formation* of a substance, $\Delta H_f°$, is the enthalpy change when one mole of the substance is produced from elements in their normal stable state at 1 atm pressure and some designated temperature, usually 25°C. For example:

$$C\ (g) + O_2\ (g) \rightarrow CO_2\ (g)$$

$$\Delta H = -393.5 \text{ kJ} = \Delta H_f° \text{ kJ/mol for } CO_2$$

$$1/2\ H_2\ (g) + 1/2\ I_2\ (s) \rightarrow HI\ (g)$$

$$\Delta H = +26.4 \text{ kJ} = \Delta H_f° = \text{kJ/mol for HI}$$

Thus, a reaction can be carried out using the two-step process below.

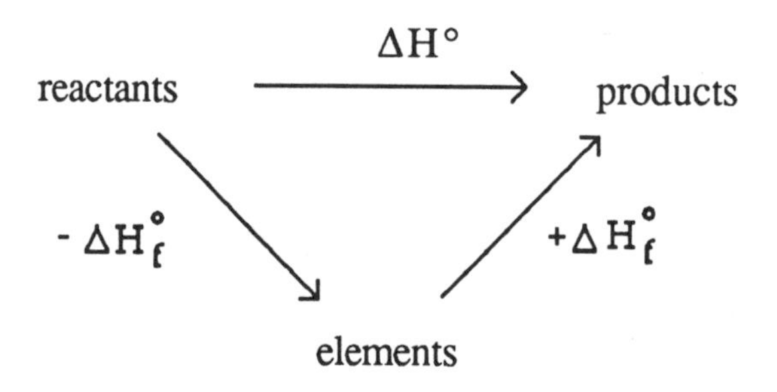

If the $\Delta H_f°$'s for each mole of products are added up and the $\Delta H_f°$'s for each mole of reactants are subtracted from this, the result is the standard enthalpy, $\Delta H°$, for this reaction. Care must be taken to note the state of each substance (*s*, *l*, *g*, *aq*, etc.) in your reaction and use the corresponding table entry. Also, the $\Delta H_f°$'s can have either positive or negative values and this must be taken carefully into account during the calculation.

Sample Problems:

Calculate the standard enthalpy change of the reaction

$$N_2H_4\,(l) + 2\,H_2O_2\,(l) \rightarrow N_2\,(g) + 4\,H_2O\,(g)$$

$\Delta H_f°$'s +50.6 -187.8 0 -241.8

$$\Delta H° = 4\,(-241.8) + 1\,(0) - 1\,(+50.6) - 2\,(-187.8) = -642.2\text{ kJ}$$

Calculate the heat of vaporization of water at 25°C.

$$H_2O\,(l) \rightarrow H_2O\,(g)$$

$\Delta H_f°$'s -285.8 -241.8

$$\Delta H° = \Delta H_{vap} = 1\,(-241.8) - 1\,(-285.8) = +\,44.0\text{ kJ}$$

Set up a sheet like the one in Worksheet 7.2 to calculate change in enthalpy from standard enthalpies of formation. This sheet is set up to handle up to four reactants and products. The formulas are entered in row 5. The coefficients from the balanced equation are entered in row 6 and the standard heats of formation in kJ/mol are entered in row 7. Type the formula =-B$6*B$7 into cell B8. Notice the minus sign in this formula. Copy this cell into C8 through E8. Type =G$6*G$7 into cell G8 and copy into cells H8 through J8. Notice in this case there was no minus sign in the formula. The final sum is calculated in cell F9 [=SUM(B8:J8)].

Solve the previous two problems using this sheet to check that it is working properly. Then use it to solve the following problems.

Calculate the standard enthalpy change for each of the following reactions. Use data from Table 7.1.

7. $3\,H_2\,(g) + N_2\,(g) \rightarrow 2\,NH_3\,(g)$

8. $2\,HI\,(g) \rightarrow H_2\,(g) + I_2\,(g)$

9. $PH_3\,(g) + 3\,HCl\,(g) + 1/2\,O_2\,(g) \rightarrow POCl_3\,(l) + 3H_2\,(g)$

Table 7.1 Standard Enthalpies of Formation (25°C)

Substance	ΔH_f° kJ/mol
C(diamond)	+1.9
C(graphite)	(0)
CH_3OH (*l*)	-238.6
CH_4 (*g*)	-74.8
C_2H_2 (*g*)	+226.7
C_2H_5OH (*l*)	-277.7
C_3H_8 (*g*)	-103.8
C_6H_6 (*l*)	+49.0
CaO (*s*)	-634.3
HCl (*g*)	-92.3
HI (*g*)	+26.4
H_2 (*g*)	(0)
H_2O (*g*)	-241.8
H_2O (*l*)	-285.8
H_2O_2 (*l*)	-187.8
I_2 (*g*)	+62.4
I_2 (s)	(0)
NH_3 (*g*)	-46.2
N_2 (*g*)	(0)
N_2H_4 (*l*)	+50.6
O_2 (*g*)	(0)
PH_3 (*g*)	+5.4
$POCl_3$ (*l*)	+542.2

Notice the values can be positive or negative and this sign must be taken into account in the calculation. Also, the values for the elements are zero since no change is taking place, forming the elements from themselves.

	A	B	C	D	E
1		Standard	Enthalpy	of	Reaction
2					
3	**Given Standard Heats of Formation**				
4		reac 1	reac 2	reac 3	reac 4
5	formula -------->	N2H4	H2O2		
6	coefficient ---->	1	2		
7	**Heat of Formation ->**	50.6	-187.8	0	0
8	(in kJ mol)	-50.6	375.6	0	0
9				**Heat =**	
10	Molecular Weight -->	32			

	F	G	H	I	J
1					
2					
3					
4		prod 1	prod 2	prod 3	prod 4
5	------>	N2	H2O		
6		1	4		
7		0	-241.8	0	0
8		0	-967.2	0	0
9	**-642.2 kJ**				
10	-20.069 kJ/gram of reac 1				

Worksheet 7.2 Sheet for Finding the Heat of a Reaction from Standard Enthalpies of Formation

Modifications

Add the following modification to calculate the heat per gram of a material (this will be the first reactant the way Worksheet 7.2 is set up). The molecular weight will be entered in cell B10 and the formula =F9/(B6*B10) is entered in cell F10.

10. Calculate the heat given off in each of the following reactions. Also calculate the heat given off per gram of material being burned in each case. Which material gives off the most heat per gram?

a) Hydrogen (MW = 2.0 g/mol)

$$H_2\ (g) + 1/2\ O_2\ (g) \rightarrow H_2O\ (l)$$

b) Methyl alcohol (MW = 32.0 g/mol)

$$CH_3OH\ (l) + 3/2\ O_2\ (g) \rightarrow CO_2\ (g) + 2\ H_2O\ (l)$$

c) Ethyl alcohol (MW = 46.0 g/mol)

$$C_2H_5OH\ (l) + 3O_2\ (g) \rightarrow 2\ CO_2\ (g) + 3\ H_2O\ (l)$$

d) Benzene (MW = 78.1 g/mol)

$$C_6H_6\ (l) + 15/2\ O_2\ (g) \rightarrow 6\ CO_2\ (g) + 3\ H_2O\ (l)$$

e) Acetylene (MW = 26.0 g/mol)

$$C_2H_2\ (g) + 5/2\ O_2\ (g) \rightarrow 2\ CO_2\ (g) + H_2O\ (l)$$

f) Methane (MW = 16.0 g/mol)

$$CH_4\ (g) + 2\ O_2\ (g) \rightarrow CO_2\ (g) + 2\ H_2O\ (l)$$

g) Propane (MW = 44.1 g/mol)

$$C_3H_8\ (g) + 5\ O_2\ (g) \rightarrow 3\ CO_2\ (g) + 4\ H_2O\ (l)$$

11. Calculate the heat given off per gram of methyl alcohol in the following reaction. You should get the same answer for kJ/g but a different answer for kJ than in part b) in the previous problem. Why?

$$2\ CH_3OH\ (l) + 3\ O_2\ (g) \rightarrow 2\ CO_2\ (g) + 4\ H_2O\ (l)$$

8

Light, Energy, and Atomic Spectra

The visible light our eyes see is just a narrow band of electromagnetic radiation. Electromagnetic radiation has a dual nature as currently seen by scientists. It can be viewed as a wave, characterized by wavelength and frequency, or it can be viewed as a stream of particles, called photons, characterized by their energy. (BLB Chap. 6)

Pictured in Figure 8.1 is a wave showing the wavelength, λ, as the distance between wave crests.

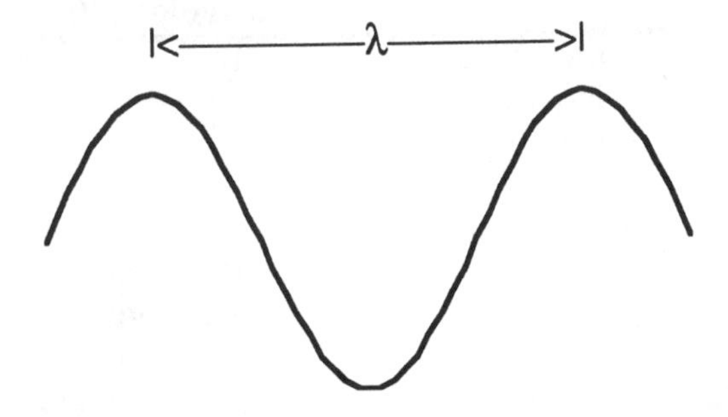

Figure 8.1 A Light Wave with a Wavelength of λ

The number of these crests that go by in a second is called the frequency, ν, and depends on the wavelength and the speed at which the light is traveling. The speed of light, c, is constant in a vacuum and equals 3×10^8 m/sec. The following equation shows the relationship between these quantities.

$$\lambda \times \nu = c$$

If the wavelength is in meters and the frequency is in sec^{-1} (hertz, Hz), then the units on the speed will be consistent with the above value in m/sec.

When light is viewed as particles, the energy of each of these particles is related to the frequency of the light when viewed as a wave.

$$E = h \times \nu$$

where h is Planck's constant and has a value of 6.63×10^{-34} J sec. The energy will then come out in joules. If you have several photons, you must multiply this energy by the total number to get the total energy.

If we combine these two equations, we can get a relationship between the energy of light and its wavelength.

$$E = h \times c / \lambda$$

You can see that the relationship is a reciprocal one: shorter wavelengths have higher energy.

Table 8.1 lists some different kinds of electromagnetic radiation and typical wavelengths.

Table 8.1 Wavelengths of Different Types of Electromagnetic Radiation

type	wavelength
radio waves	100 m
microwaves	1 cm
infrared	3 μm
visible light	500 nm
ultraviolet	100 nm
x-rays	100 pm
gamma rays	0.1 pm

Let's calculate the frequency of some electromagnetic radiation that has a wavelength of 500 nm. Rearrange the equation relating frequency to wavelength and change the wavelength units to meters to give:

$$\nu = c / \lambda = 3 \times 10^{8} / 5 \times 10^{-7} = 6 \times 10^{14} \text{ sec}^{-1}$$

This 500 nm (5×10^{-7} m) radiation is in the visible region of the spectrum.

Now let's calculate the energy of this light from the wavelength.

$$E = 6.63 \times 10^{-34} \times 3 \times 10^{8} / 5 \times 10^{-7} = 3.98 \times 10^{-19} \text{ J}$$

This is the energy for one photon with this 500-nm wavelength. For one mole of photons, multiply by Avogadro's number (6.02×10^{23}) to get 239000 J/mol or 239 kJ/mol.

Set up a sheet as shown in Worksheet 8.1 that will calculate wavelength from frequency. Input the frequency in cell B3 and put the formula for wavelength in cell B4 (=3E8/B3). Use the numbers in the example above to check your sheet and then use it to work the following problem.

1. Use your sheet to calculate the wavelength of the following radio stations:

KJRB (AM, Spokane, WA) 790 kHz (790,000 Hz)
KDRK (FM, Spokane, WA) 93.7 MHz (93.7x10^6 Hz)

Your favorite AM station
Your favorite FM station

Do FM or AM stations use the longer wavelength? Be careful about the units your sheet uses. If data is entered with incorrect units, the answer will be incorrect.

Now set up a sheet that will calculate frequency, the energy of one photon (in J), and the energy of a mole of photons (in kJ/mol) from the wavelength in meters as in Worksheet 8.2. Input the wavelength (in m) in cell C3. Calculate the frequency (Hz) in cell C4 (=3E8/C3). Calculate the energy of one photon in cell C5 (=6.63E-34*C4). Calculate the energy (in kJ) for a mole in cell C6 (=6.02E23*C5/1000). Use the previous example to check the functioning of your sheet. Then copy cells C3 to C6 into cells D3 to I6. If you add labels and put in the wavelengths from the previous table, you should end up with Worksheet 8.2. Use the results of this sheet to consider the following problem:

2. Calculate the energy of radio waves (100 m) and x-rays (100 pm) and compare these values to the bond energy of a typical chemical bond (400kJ/mol). Why do you think x-rays are hazardous to us, while we are continually bombarded with radio waves with no apparent harm?

	A	B
1	**Electromagnetic**	**Radiation**
2		
3	frequency (Hz)	6E+14
4	wavelength (m)	0.0000005

Worksheet 8.1 Sheet for Calculating Wavelength from Frequency

	A	B	C	D
1	Electromagnetic	Radiation		
2			radio	microwave
3	wavelength	m	100	0.01
4	frequency	Hz	3000000	3E+10
5	energy	J	1.989E-27	1.989E-23
6	energy/mole	kJ/mol	1.197E-06	0.0119738

	E	F	G	H	I
1					
2	infrared	visible	uv	x-rays	gamma
3	0.000003	0.0000005	0.0000001	1E-10	1E-13
4	1E+14	6E+14	3E+15	3E+18	3E+21
5	6.63E-20	3.978E-19	1.989E-18	1.989E-15	1.989E-12
6	39.9126	239.4756	1197.378	1197378	1.197E+09

Worksheet 8.2 Sheet for Calculating Frequency and Energy from Wavelength

The Hydrogen Atom and Atomic Spectra

A neutral hydrogen atom has one electron moving about its nucleus. Its negative charge balances the positive charge of the one proton in the nucleus. This electron can absorb light and gain energy or it can lose energy and emit light in the process. So the energy of the electron can vary. At the turn of the century, scientists realized that the energy of this electron could not take on any arbitrary value even though it could change. This was determined from looking at the wavelengths or different colors of light that were emitted by hydrogen atoms. Only a few specific wavelengths or colors could be observed instead of all possible ones. This connection between light and the energy of the electron was found to be the following.

If a hydrogen atom is excited, i.e., its electron has its energy raised above the minimum or most stable value, the electron will eventually fall in energy. The energy it loses cannot just disappear. It is converted to a photon or light particle whose energy is equal to the energy lost by the electron. Thus energy is conserved. The frequency of the light is related to its energy by the equation $E = h\,\nu$, and the wavelength, and thus the color, is related to the frequency by $\lambda = c/\nu$. Thus the change in energy of the electron is directly related to the color of the light emitted.

$$\Delta E_{\text{electron}} = E_{\text{light}} = h\,c\,/\,\lambda$$

where $h = 6.63 \times 10^{-34}$ J sec (Planck's constant) and $c = 3 \times 10^{8}$ m/sec (the velocity of light).

Thus if only certain colors are observed there must be restrictions on the separations between the electron's energy levels in the hydrogen atom.

The Bohr model of the hydrogen atom leads to an expression for the allowed energy values or energy levels for the electron. These values are described in the equation:

$$E = -2.18 \times 10^{-18} (1/n^2) \text{ joules}$$

where n can take on any integer value (1, 2, 3, ... etc.). The fact that n is an integer limits the values E can have and the electron is restricted to jumping between these levels as it changes energy.

The above equation can be simply modified to also handle hydrogen-like ions. These are ions of other elements that contain a total of only one electron just like hydrogen, for example He^{+} and Li^{++}. The modified equation just includes the nuclear charge or atomic number, z, and reduces to the above equation for hydrogen where $z = 1$.

$$E = -2.18 \times 10^{-18} (z^2/n^2) \text{ joules}$$

This equation and the Bohr model of the atom will not work for neutral atoms of other elements as they all contain more than one electron. Even though this equation will not exactly describe the other elements, some of the trends we will note for the hydrogen-like ions will qualitatively fit the other elements.

What is the energy of an electron at the $n = 3$ level in the He^{+} ion?

$$E = -2.18 \times 10^{-18} \times 2^2/3^2 = -9.69 \times 10^{-19} \text{ J}$$

Notice that the energy values will be negative and because of this the lowest value will be when $n = 1$. The highest value will be when $n = \infty$ with $E = 0$. This corresponds to the electron having enough energy to escape the attraction of the nucleus.

Set up a sheet like the one in Worksheet 8.3. We will input the nuclear charge in cell B3. Ten different quantum levels are entered in cells B4 through B13. Put the formula for the energy in cell D4.

```
=-2.18E-18*$B$3^2/B4^2
```

Copy this cell into cells D5 through D13. To see if your sheet is working correctly, input z=2 and check with the value of level 3 that you calculated above.

	A	B	C	D	E
1	Energy of H-like ions				
2					
3	charge,z	2		energy	joules
4	level, n	1		-8.72E-18	
5		2		-2.18E-18	
6		3		-9.68889E-19	
7		4		-5.45E-19	
8		5		-3.488E-19	
9		6		-2.42222E-19	
10		7		-1.77959E-19	
11		8		-1.3625E-19	
12		9		-1.07654E-19	
13		10		-8.72E-20	

Worksheet 8.3 Sheet to Calculate Energy in Hydrogen-like Ions

Now work the following problem using this sheet.

1. Calculate the first five energy levels (n = 1, 2, 3, 4, and 5) for H (atomic number = 1). Calculate the first ten levels for He^{+} (n = 1 to 10, atomic number = 2). Sketch the two sets of energy levels vertically side by side and compare them. Use a whole sheet of paper because the upper levels will be very close together. Both sketches should have E = 0 included at the top of your paper. Be sure to label each level with its n value and its energy value. How much does increasing the nuclear charge lower the bottom level in He^{+} compared to H?

Now modify your sheet to look like Worksheet 8.4.

	A	B	C	D	E	F
1	Energy change in H-like ions				energy	wave-
2	nuclear				change	length
3	charge,z	1	n1	n2	(joules)	(nm)
4			3	2	-3.028E-19	656.91743
5			4	2	-4.088E-19	486.6055
6			5	2	-4.578E-19	434.4692
7			6	2	-4.844E-19	410.57339
8			7	2	-5.005E-19	397.3945
9			8	2	-5.109E-19	389.2844

Worksheet 8.4. Sheet to Calculate Energy Change and Radiation Wavelength with Electron Transitions

Again the nuclear charge is entered in cell B3. The n value for the energy level that the electron is on initially goes in cell C4. The final level is entered in cell D4. The energy change is calculated in cell E4.

=-2.18E-18*$B$3^2*(1/D4^2-1/C4^2)

The wavelength (in nm) of the light emitted or absorbed is calculated in cell F4.

=ABS(6.63E-34*3E17/E4)

The absolute value function makes sure the wavelength comes out positive even if the energy change is negative, as will be the case if the electron is falling from a higher to a lower level. Now copy cells E4 and F4 into cells E5 through F9 so you can do six calculations at one time. Your sheet should look like Worksheet 8.4 if you enter the same data.

2. Calculate the wavelength of light given off by H when the electron falls between the following energy levels:

$6 \rightarrow 2$
$5 \rightarrow 2$
$4 \rightarrow 2$
$3 \rightarrow 2$

Compare these numbers to the observed lines in the visible H emission spectrum: 411nm, 434 nm, 487 nm, and 657 nm. Show and label these transitions on your previous energy level diagram for H.

3. By trial and error, find which transitions lead to emission of visible light from He^+ (400 - 700 nm). Trial and error means calculating the wavelength for several pairs of levels (with nuclear charge = 2) and seeing which fall in the range for visible light (400 to 700 nm). Show and label these transitions on your previous energy level diagram for He^+.

4. Neutral hydrogen is observed in interstellar space by detecting 21 cm radiation using radio telescopes. Between what two levels is the electron falling in hydrogen when it emits this radiation? By trial and error, find the two levels next to each other (such as 138 and 137) that give this wavelength.

5. Calculate the ionization energy of H and He^{+}. This corresponds to the electron going from the ground state (n = 1) to escaping from the nucleus ($n = \infty$). Use $n = 1 \times 10^6$ for ∞. Does a higher nuclear charge increase or decrease the ionization energy?

9

Atomic Orbitals

Atomic orbitals are plots that describe the probability of finding the electrons in an atom at different points in three-dimensional space around the nucleus. They can be looked at as telling us, on the average, where the electrons are, or what the electron density is around an atom. These plots are calculated from functions that are solutions to quantum mechanical equations describing atoms. (BLB Chap. 6)

Orbitals have size, shape, and orientation associated with them. These attributes, along with the energy of the electrons and the number of electrons in the outer orbitals, determine much about how an atom forms chemical bonds. The size of the orbitals is perhaps the most fundamental property.

The sizes of the outer orbitals of an atom determine the size of the atom itself and, along with the nuclear charge, determines how tightly the outer, bonding electrons are held. Electronegativity is a measure of this attraction for electrons in a chemical bond. The relative electronegativities of the atoms involved in a bond determine which of the three main types of bonding these atoms will use: metallic, ionic, or covalent.

We will study what affects the size of orbitals by plotting some of them. This will lead to an understanding of why the size trends of atoms in the periodic table exist as they do, with atoms becoming smaller as we move across the table but larger as we go down. We will also study to some extent the shapes of the more common orbitals.

The following spreadsheets plot two-dimensional x-z (with z vertical and x horizontal) cross-sections of several orbitals. The sheets plot a character if the probability of the electron being at that point is above a certain level. It leaves a blank otherwise. The plots will show the overall shape and relative size of each orbital so you can make comparisons.

Technically the plots are made for hydrogen atoms and hydrogen-like (one electron) ions, but these will show the trends in multi-electron atoms also. A random number is included in the cutoff point of the plot to help realistically represent the probability nature of these plots. This leads to ragged edges around the orbitals and a slightly different plot each time the same orbital is run. This shows the uncertainty of the electron's position. The overall size, shape, and orientation will still be the same, however.

Worksheet 9.1 shows the start of a spreadsheet for plotting atomic orbitals. The nuclear charge will be entered in cell B3. Enter the formula =$B$3^1.5 in cell B4. This raises the charge to the 1.5 power. The number in cell B5 (0.529) is a constant used in the calculation. The x coordinates of our plot are placed in row 1 starting with -3.3 in cell D1. The formula =D1+.33 is placed in cell E1 and copied into cells F1 through X1. The z coordinates are started at -3.3 in cell C2 with the formula =C2+0.33 placed into cell C3 and copied into cells C4 through C22.

Worksheet 9.2 shows the columns for the plot reduced in width and the formula for the 2s orbital entered. Select columns C through X by dragging the mouse over the letters with the mouse button held down. Then select Format, Column Width, from the menu bar. Enter the number 1.57 in the column width box and click on OK. Next type the following formula into cell D2. This formula calculates the probability of finding the electron at the x and z coordinates in cells D1 and C2. If the value is equal to or greater than 0.05, a * is printed; otherwise the cell is left blank. Be very careful in entering this formula since it is fairly complex.

```
=IF(RAND()*(1.4*$B$4*(2-$B$3*SQRT(D$1^2+$C2^2)/$B$5)
*EXP(-$B$3*SQRT(D$1^2+$C2^2)/(2*$B$5)))^2>=0.05,"*"," ")
```

Now copy this formula into all the cells from D2 to X22. The sheet will then start plotting the orbital in this square area and should end up looking similar to Worksheet 9.2. Type a + into cell N12 if you wish to show the position of the nucleus. Your plot will not necessarily be exactly like the example because of the random function placed in the formula, but it should have the same overall shape and size.

1. The plot you have just made is for a hydrogen atom. Now change the charge to 1.3 and the plot will be redone. It may not look a lot different than the first. Next try 1.6, 1.8, and finally 2.0. This last value corresponds to the He^{+} ion. Try it with a charge of 3. This corresponds to the Li^{++} ion. Has the overall size of the orbital changed as the nuclear charge has increased? What about the shape?

	A	B	C	D	E	F	G
1	Hydrogen-like	Orbitals		-3.3	-2.97	-2.64	-2.31
2	2s		-3.3				
3	charge -->	1	-2.97				
4		1	-2.64				
5		0.529	-2.31				
6			-1.98				
7			-1.65				
8			-1.32				
9			-0.99				
10			-0.66				

Worksheet 9.1 Starting a Spreadsheet for Plotting Atomic Orbitals

	A	B	C	D	E	F	G	H	I	J	K	L	M	N	O	P	Q	R	S	T	U	V	W	X
1	Hydrogen-like	Orbitals		-3	-3	-3	-2	-2	-2	-1	-1	-1	0	0	0	1	1	1	2	2	2	3	3	3
2	2s		-3											*	*									
3	charge -->	1	-3						*					*	*		*							
4		1	-3				*			*			*	*	*	*	*							
5		0.529	-2						*			*		*	*	*	*		*		*			
6			-2			*				*	*		*	*	*		*		*	*				
7			-2		*	*	*		*							*	*	*			*	*		
8			-1			*	*		*								*	*			*	*		
9			-1					*		*								*		*	*	*		
10			-1		*			*	*				*	*					*		*	*		
11			0		*	*	*		*			*	*	*	*	*				*	*		*	
12			0			*	*	*				*	*	+	*	*					*		*	
13			0		*	*	*					*	*	*	*	*			*	*	*			*
14			1			*	*			*			*		*			*			*			
15			1			*		*	*	*									*	*			*	
16			1					*			*						*	*		*				
17			2		*		*					*		*	*			*				*		
18			2			*	*	*	*		*		*			*		*	*	*	*			
19			2									*		*	*			*	*					
20			3				*			*							*			*				
21			3														*		*					
22			3								*													

Worksheet 9.2 Spreadsheet for Plotting the Hydrogen 2s Orbital

Now let's change the function in cell D2 to the following:

```
=IF(RAND()*($B$4*($B$3*SQRT(D$1^2+C$2^2)/$B$5)*EXP(-
$B$3*SQRT(D$1^2+$C2^2)/(2*$B$5))*($C2/(SQRT(D$1^2+$C
2^2))))^2>=0.05,"*"," ")
```

Copy this from cells D2 through X22. Again put a + into cell N12 to show the position of the nucleus. The $2p_z$ orbital is now plotted and should look similar to Worksheet 9.3 if you change the charge back to 1.

2. Next run the $2p_z$ orbital with the nuclear charge equal to 1.3, 1.6, 1.8, and 2.0 (this last value corresponds to the He^+ ion). Try it with a charge of 3. This corresponds to the Li^{++} ion. How does the increased nuclear charge affect the size of the $2p_z$ orbital? Is this the same trend as for the 2s orbital?

Now let's change the function in cell D2 to the following:

```
=IF(RAND()*(1.4*$B$4*EXP(-$B$3*SQRT(D$1^2+$C2^2)
/$B$5))^2>=0.05,"*"," ")
```

Copy this from cells D2 through X22. Again put a + into cell N12 to show the position of the nucleus. The 1s orbital is now plotted and should look similar to Worksheet 9.4 if you change the charge back to 1.

3. Note that the 2s and $2p_z$ orbitals you ran first were about the same size with a charge of 1. Compare the size of these two orbitals to the 1s orbital that was run with a nuclear charge of 1. On what does size depend (assuming the same charge): the quantum number *n*, which determines energy, or the quantum numbers *l* and *m*, which determine shape and orientation of the orbitals? You can run the 1s orbital with different charges to help decide this, for example use 1.3, 1.7, and 2 for the charge. Try it with a charge of 3. This corresponds to the Li^{++} ion.

Now let's change the function in cell D2 to the following:

```
=IF(RAND()*($B$4*($B$3*SQRT(D$1^2+$C2^2)/$B$5)*EXP(-
$B$3*SQRT((D$1^2+$C2^2)/(2*$B$5))*(D$1/(SQRT(D$1^2+$C
2^2))))^2>=0.05,"*"," ")
```

Copy this from cells D2 through X22. Again put a + into cell N12 to show the position of the nucleus. The $2p_x$ orbital is now plotted and should look similar to Worksheet 9.5 if you change the charge back to 1.

	A	B	C	D	E	F	G	H	I	J	K	L	M	N	O	P	Q	R	S	T	U	V	W	X
1	Hydrogen-like	Orbitals		-3	-3	-3	-2	-2	-2	-1	-1	-1	0	0	0	1	1	1	2	2	2	3	3	3
2	2pz		-3								*				*									
3	charge -->1		-3									*	*	*	*			*						
4		1	-3								*		*	*	*	*	*	*	*					
5		0.529	-2							*	*	*	*	*	*	*								
6			-2						*			*	*	*	*	*			*					
7			-2						*	*		*	*	*		*	*							
8			-1					*			*	*	*	*	*		*	*	*					
9			-1						*	*	*	*	*	*	*	*	*	*	*					
10			-1							*		*	*	*		*	*	*						
11			0									*	*	*	*									
12			0											+										
13			0										*	*		*								
14			1									*	*	*			*		*					
15			1						*			*	*	*	*	*	*		*					
16			1						*	*		*	*	*	*	*		*						
17			2					*		*	*	*	*	*	*	*	*	*	*					
18			2						*	*	*		*	*	*	*		*						
19			2						*		*		*		*		*		*					
20			3								*	*	*	*	*	*	*	*						
21			3							*		*			*	*	*							
22			3											*										

Worksheet 9.3 Spreadsheet for Plotting the $2p_z$ Orbital

	A	B	C	D	E	F	G	H	I	J	K	L	M	N	O	P	Q	R	S	T	U	V	W	X
1	Hydrogen-like	Orbitals		-3	-3	-3	-2	-2	-2	-1	-1	-1	0	0	0	1	1	1	2	2	2	3	3	3
2	1s		-3																					
3	charge -->1		-3																					
4		1	-3																					
5		0.529	-2																					
6			-2																					
7			-2																					
8			-1																					
9			-1																					
10			-1											*										
11			0										*	*	*									
12			0									*	*	+	*	*								
13			0										*	*	*	*								
14			1										*		*									
15			1																					
16			1																					
17			2																					
18			2																					
19			2																					
20			3																					
21			3																					
22			3																					

Worksheet 9.4 Spreadsheet for Plotting the 1s Orbital

	A	B	C	D	E	F	G	H	I	J	K	L	M	N	O	P	Q	R	S	T	U	V	W	X
1	Hydrogen-like	Orbitals		-3	-3	-3	-2	-2	-2	-1	-1	-1	0	0	0	1	1	1	2	2	2	3	3	3
2	2px		-3																					
3	charge -->	1	-3																					
4		1	-3																					
5		0.529	-2																					
6			-2																					
7			-2						*								*	*	*	*				
8			-1				*	*		*							*	*		*	*			
9			-1			*	*		*	*	*	*				*	*	*		*	*	*	*	
10			-1		*	*	*	*	*	*	*	*					*	*	*	*				
11			0			*	*		*	*	*	*	*		*	*	*	*	*	*	*	*		*
12			0		*			*	*	*			*	+	*	*	*	*	*	*	*		*	
13			0	*	*		*	*	*	*	*		*		*	*	*		*		*		*	*
14			1		*		*	*			*	*			*	*	*	*	*	*		*	*	*
15			1			*	*	*	*	*	*					*	*			*	*			
16			1				*		*							*		*	*	*	*	*		
17			2						*		*						*	*	*	*				
18			2															*		*	*			
19			2																					
20			3																					
21			3																					
22			3																					

Worksheet 9.5 Spreadsheet for Plotting the $2p_x$ Orbital

4. The s and p designations refer to the value of the l quantum number with $l=0$ for s and $l=1$ for p. Are the 1s and 2s orbitals alike in size or shape? How are the $2p_x$ and $2p_z$ orbitals alike? Do you think the l quantum number determines size, shape, or orientation of the atomic orbitals? Does the $2p_x$ orbital behave like the others when the charge is changed? Plot it for several charges between 1 and 2 to find out. Try it with a charge of 3. This corresponds to the Li^{++} ion.

In multi-electron atoms, the inner electrons screen the nucleus so that the outer electrons only see part of the total nuclear charge. This effective nuclear charge can be used in our program to see the effects on orbital size as we go through the elements in the periodic table. We have seen above that the quantum number, *n*, also affects the size. These two factors are at play in determining the size trends of atoms in the periodic table. We will study these trends in the next two problems.

5. Run the $2p_z$ orbital three times using 2.60, 3.90, and 5.20 for the nuclear charges. These correspond to the charges that the outer 2p electrons in boron, nitrogen, and fluorine see after the nucleus has been

screened by the 1s and 2s electrons. The atomic numbers and electronic configurations of these elements are:

B	5	$1s^22s^22p^1$
N	7	$1s^22s^22p^3$
F	9	$1s^22s^22p^5$

Notice as we go across the periodic table that the atomic number (or actual nuclear charge) of these elements increases. The effective nuclear charge also increases. How does the size of the outer orbitals change as the charge increases as we go across the table?

6. Run the 1s orbital with a nuclear charge of 1. This is the outer orbital containing an electron in H since the electronic configuration for H is $1s^1$. Next run the 2s orbital with an effective nuclear charge of 1.30. This represents the 2s orbital for Li ($1s^22s^1$) with the nucleus having a charge of +3 being screened by the 2 electrons in the 1s orbital so that that the outer 2s electron sees only a charge of 1.30. Lithium is directly under hydrogen in the periodic table and this allows us to see the effects of changing both the nuclear charge and the quantum number, *n*, on the size of the outer orbitals and thus on the size of the atoms as we go down the periodic table. Which effect is larger when going down the periodic table: an increased nuclear charge making the orbital smaller or an increased *n* value making the orbital larger, i.e., do the atoms get smaller or larger as we go down the periodic table?

The two trends in problems 5 and 6 hold in most cases throughout the whole periodic table, explaining why the smaller, more electronegative atoms are in the upper right-hand corner and the larger, less electronegative atoms are in the lower left-hand corner. Looking at it another way, the larger metal atoms tend to be on the left side and the smaller nonmetal atoms are on the right side.

10

Molecular and Hybrid Orbitals

When a covalent chemical bond forms, the shared electrons are able to travel back and forth between the two atoms involved because of the formation of molecular orbitals that extend simultaneously over the two atoms. In this chapter, we will study the formation of these orbitals and thus how chemical bonds form by plotting the probability of finding electrons in different positions around the atoms. (BLB Chap. 9)

Covalent bonds are formed between two bonding atoms if they both are small and have a strong attraction for electrons or have a high electronegativity. Thus, it is nonmetals that bond to each other in this manner. (Metallic bonds form between metal atoms and ionic bonds form between a metal and a nonmetal.) Since both atoms attract the electrons strongly, the atoms engage in an eternal tug of war, sharing the electrons between them.

For this sharing to take place the original atomic orbitals that restricted the electrons to one atom or the other must be replaced. The new orbitals must extend over both atoms at the same time for sharing to be possible. The quantum mechanical calculations needed to describe these orbitals exactly are very long and complicated. However, a good approximation of these can be obtained from what is called linear combination of atomic orbitals-molecular orbitals theory (LCAO-MO).

This is a complicated sounding name, but it is easy to visualize what is happening when a bond forms because we will start with the usual atomic orbitals with which we are already familiar. Imagine the two atomic orbitals, each containing one bonding electron, overlapping more and more as the two atoms approach each other. Now visualize the total electron density of these two overlapped atomic orbitals as the final total electron density of the two electrons being shared between the bonding atoms.

The LCAO-MO theory does just this by adding the two functions together that describe the atomic orbitals on each atom and using the resulting new function to calculate the probability or electron density around the two atoms. This gives a function that describes the electron probability in a bond reasonably well.

Two s orbitals, two p orbitals end to end, or an s and the end of a p orbital can be overlapped to form molecular orbitals. Which type to use depends on which atomic orbitals contain the bonding electrons. These resulting three types of molecular orbitals are called *sigma* orbitals because they all look like an s orbital when viewed from the end of the bond. Two p orbitals can also be overlapped side to side to form a *pi* molecular orbital. This pi molecular orbital looks just like a p orbital when viewed from the end.

We will set up a sheet that will plot some of these orbitals. The functions used to describe the molecular orbitals are quite long and will not fit into one cell on our sheet, so we will set up what is called a *function macro* sheet first that will describe each of the atomic orbitals we wish to use. We will then combine these on our sheet that plots the molecular orbital. Look at Worksheet 10.1M to see how this macro sheet will look.

To start a macro worksheet, select File, New..., Macro Sheet, and OK. Each column of Worksheet 10.1M contains the cell information that will describe one of the orbitals. These are labeled at the head of each column. Enter pz1 into cell A1. To use this as the name of our function, select this cell and click on Formula, Define Name..., and the Macro Function circle to darken it. Then click on the Name box and enter the name pz1. The Refers to: box should be =$A$1. Then click on OK. Now type =ARGUMENT("x") in cell A3. This will transfer the *x* coordinate to our function.

Type similar statements into cells A4 and A5 for "z" and "d." These are the *z* coordinate and distance between the two atoms. A constant equal to the radius of the first orbit of an electron in the Bohr model of the hydrogen atom is entered in cell A6 as 0.529 (Å). Define this cell A6 as "a" using Formula, Define Name as done above, except do not darken the Macro Function circle. This name will be used as shown in the function below. The function for a $2p_z$ orbital (z axis vertical) is entered into cell A7:

```
=0.8*(SQRT((x+d/2)^2+z^2)/a*EXP(-SQRT((x+d/2)^2+z^2)/(2*a)))
        *z/SQRT((x+d/2)^2+z^2)
```

Notice that unlike on a regular sheet, which normally displays values, the macro sheet displays the formula. Next enter =RETURN(A7) in cell A8. Finally, save this sheet on disk as O (the letter) using File, Save As..., entering the name O in the box, and clicking on OK. It will be saved as O.XLM on the disk but the extender XLM will be added for you by Excel.

	A
1	pz1
2	
3	=ARGUMENT("x")
4	=ARGUMENT("z")
5	=ARGUMENT("d")
6	0.529
7	=0.8*(SQRT((x+d/2)^2+z^2)/a*EXP(-SQRT((x+d/2)^2+z^2)/(2*a)))*z/SQRT((x+d/2)^2+z^2)
8	=RETURN(A7)

	B
1	pz2
2	
3	=ARGUMENT("x")
4	=ARGUMENT("z")
5	=ARGUMENT("d")
6	0.529
7	=0.8*(SQRT((x-d/2)^2+z^2)/a*EXP(-SQRT((x-d/2)^2+z^2)/(2*a)))*z/SQRT((x-d/2)^2+z^2)
8	=RETURN(B7)

Worksheet 10.1M Macro Sheet for Defining Atomic Orbital Functions

Now copy cells A1 through A8 to cells B1 through B8. Change cell B2 to pz2 and go through the define name procedure for defining the macro function name pz2 as referring to cell B1. Change cell B7 by editing the expression now there by changing +d to -d in all three locations in this cell. This defines a $2p_z$ orbital on the second atom involved in the bond formation. Cell B8 should read =RETURN(B7) after you copied it. The other cells remain unchanged. Again, save this sheet as O.

NOTE: *This function macro sheet, O.XLM, must always be opened before using any of the plotting sheets in this chapter so the functions are available to these sheets. Do not close this sheet before setting up the following sheet. If you are working on this chapter at a later time, be sure to open O.XLM from disk before using any of the plotting sheets.*

Before we add the other orbitals needed later to this sheet, let's set up the sheet to plot orbitals and try it out with the two orbitals we have just defined. Worksheet 10.2 shows such a sheet. Start by selecting columns A through AI by dragging the mouse cursor across the column heading holding down the left mouse button. Then select Format, Column Width..., enter 1.88 in the box, and click on OK. This narrows the column width to the same size as the row height. Next select cell A3 and define its name as D selecting Formula, Define Name..., entering D in the Name box, and clicking on OK.

Enter -5.28 in cell C1 and =C1+0.33 in cell D1. Then copy D1 into cells F1 through AI1. Enter -3.3 in cell B2 and =B2+0.33 in cell B3. Copy cell B3 into cells B4 through B22. This sets up an *x*-*z* coordinate system with *z* vertical and *x* horizontal similar to that in Chapter 9 used for atomic orbital plots. The function for plotting a pi molecular orbital by overlapping two vertical p orbitals is entered into cell C2:

```
=IF((O.XLM!pz1(C$1,$B2,D)+O.XLM!pz2(C$1,$B2,D))^2>0.1,"*"," ")
```

The two functions we defined on the macro sheet above are added together and squared, and; if the resulting probability is above a certain level, an * character is plotted, otherwise the cell is left blank. Be careful about upper and lower case in this formula. C$1 is the *x* coordinate, $B2 is the *z* coordinate, and D is the distance between the two atoms. Now enter our first distance in cell A3. We will use the value of 7 so the two atoms are far enough apart that the two p orbitals are seen separated from each other. Finally, copy the formula in cell C2 into cells C2 through AI22 and the calculation and "plotting" will start. These are lengthy calculations and you may have to wait awhile for the finish, so be patient! If everything went well, your sheet looks like Worksheet 10.2.

	A	B	C	D	E	F	G	H	I	J	K	L	M	N	O	P	Q	R	S	T	U	V	W	X	Y	Z	AA	AB	AC	AD	AE	AF	AG	AH	AI
1			-5	-5	-5	-4	-4	-4	-3	-3	-3	-2	-2	-2	-1	-1	-1	0	0	0	1	1	1	2	2	2	3	3	3	4	4	4	5	5	5
2		-3																																	
3	7	-3																																	
4		-3					*	*	*	*																		*	*	*	*				
5		-2				*	*	*	*	*	*																*	*	*	*	*	*			
6		-2			*	*	*	*	*	*	*	*														*	*	*	*	*	*	*	*		
7		-2			*	*	*	*	*	*	*	*														*	*	*	*	*	*	*	*		
8		-1			*	*	*	*	*	*	*	*														*	*	*	*	*	*	*	*		
9		-1			*	*	*	*	*	*	*	*														*	*	*	*	*	*	*	*		
10		-1				*	*	*	*	*	*																*	*	*	*	*	*			
11		0						*	*																				*	*					
12		0																																	
13		0						*	*																				*	*					
14		1				*	*	*	*	*	*																*	*	*	*	*	*			
15		1			*	*	*	*	*	*	*	*														*	*	*	*	*	*	*	*		
16		1			*	*	*	*	*	*	*	*														*	*	*	*	*	*	*	*		
17		2			*	*	*	*	*	*	*	*														*	*	*	*	*	*	*	*		
18		2			*	*	*	*	*	*	*	*														*	*	*	*	*	*	*	*		
19		2				*	*	*	*	*	*																*	*	*	*	*	*			
20		3					*	*	*	*																		*	*	*	*				
21		3																																	
22		3																																	

Worksheet 10.2 Sheet Showing Two p Orbitals on Two Atoms Some Distance Apart

Make sure your sheet is working correctly before continuing to the following exercises.

You will now observe what happens to the electron distribution as you bring the two atoms together to form a pi bonding molecular orbital.

1. Run the sheet again with a 5 Å separation and note how the two orbitals start to distort, almost like they are attracting each other.

2. Run the sheet one more time with a distance of 3.5 Å and note how this distortion effect increases with the electron density building up between the two atoms leading to the overlap of the two atomic orbitals.

3. Run again using 1.5 Å. This shows how the two p orbitals have coalesced into one molecular orbital with the 2 electrons being able to pass back and forth between the 2 nucleii forming a covalent bond.

To see how sigma type orbitals look, you need to add additional orbital functions to your function macro sheet. Then you can make some simple changes in your plotting sheet and rerun it again to see additional types of molecular orbitals.

Back on your macro sheet, copy cells A1 through A8 into cells C1 through G8. You will then need to change the names in cells C1 though G1 to those shown in Worksheet 10.3M. These names need to be defined as you did before. The only other changes you need to make are in row 7 to define the other atomic orbitals. These are listed by cell location below.

```
C7: =0.8*(SQRT((x+d/2)^2+z^2)/a*EXP(-SQRT((x+d/2)^2+z^2)/(2*a)))
        *(x+d/2)/SQRT((x+d/2)^2+z^2)
```

This is almost identical to cell A7, so you could copy cell A7 into C7 and edit it to save some typing.

```
D7: =0.8*(SQRT((x-d/2)^2+z^2)/a*EXP(-SQRT((x-d/2)^2+z^2)/(2*a)))
        *(x-d/2)/SQRT((x-d/2)^2+z^2)
```

This is identical to cell C7 except +d is changed to -d.

```
E7: =4.5*EXP(-SQRT((x+d/2)^2+z^2)/a)
```

```
F7: =4.5*EXP(-SQRT((x-d/2)^2+z^2)/a)
```

This is identical to E7 except +d is changed to -d.

```
G7: =1.4*(2-SQRT((x+d/2)^2+z^2)/a)*EXP(-SQRT((x+d/2)^2+z^2)
        /(2*a))
```

	C	D	E	F	G
1	px1	px2	xs1	xs2	x2s1
2					
3	=ARGUMENT("x")	=ARGUMENT("x")	=ARGUMENT("x")	=ARGUMENT("x")	=ARGUMENT("x")
4	=ARGUMENT("z")	=ARGUMENT("z")	=ARGUMENT("z")	=ARGUMENT("z")	=ARGUMENT("z")
5	=ARGUMENT("d")	=ARGUMENT("d")	=ARGUMENT("d")	=ARGUMENT("d")	=ARGUMENT("d")
6	0.529	0.529	0.529	0.529	0.529
7	=0.8*(SQRT((x+d/2)	=0.8*(SQRT((x-d/2)	=4.5*EXP(-SQRT((x	=4.5*EXP(-SQRT((x	=1.4*(2-SQRT((x+d/
8	=RETURN(C7)	=RETURN(D7)	=RETURN(E7)	=RETURN(F7)	=RETURN(G7)

Worksheet 10.3M Remaining Orbitals Defined on the Function Macro Sheet

Again save this macro sheet as O. Next modify your plotting sheet to see some sigma type bonding molecular orbitals. Change cell C2 to:

=IF((O.XLM!xs1(C$1,$B2,D)+O.XLM!xs2(C$1,$B2,D))^2>0.1,"*"," ")

This differs only slightly from what is already in that cell. Just change pz to xs in two places and copy this cell from C2 through AI22. The bond that you will see forming is an s-s sigma bond, typical of the one found in H_2. Run the sheet using 7, 5, 3.0, and then 1.5 Å.

For s-p sigma bonds such as found in HF, change cell C2 to the following and copy into C2 through AI22:

=IF((O.XLM!xs1(C$1,$B2,D)+O.XLM
!px2(C$1,$B2,D))^2>0.1,"*"," ")

This differs from what is already there by changing xs2 to px2. Again copy this cell into C2 through AI22. Run using 7, 5, 3.5, and 1.5 Å.

For a p-p sigma bond such as found in F_2, change cell C2 to the following and copy into the others:

=IF((O.XLM!px1(C$1,$B2,D)+O.XLM
!px2(C$1,$B2,D))^2>0.1,"*"," ")

You only need to change xs to px to make this change. Run at 8, 6, 4, and 2 Å.

Notice that in all these cases the electron density between the two bonding atoms increases. This is typical of bond formation when molecular orbitals are formed and electrons are shared between two atoms.

Hybrid Atomic Orbitals

In many cases, the geometry or arrangement of the atomic orbitals around a bonding atom is not in agreement with the actual molecular shape or geometry. One way to handle this situation is to form *hybrid* atomic orbitals. These are orbitals that are made by mathematically combining the usual s and p orbitals to get a new set of orbitals that point in the right direction so that the shape of the molecule will be correct. These hybrid orbitals can then be overlapped with atomic or other hybrid atomic orbitals on neighboring atoms to form molecular orbitals.

Among the most common types used are:

sp (made from one s and one p)
2 orbitals in a linear arrangement 180° apart.

sp^2 (from one s and two p's)
3 orbitals in a trigonal arrangement 120° apart.

sp^3 (from one s and three p's)
4 orbitals in a tetrahedral arrangement 109° apart.

Our molecular orbital plotting sheet can be easily modified to plot some of these hybrid orbitals. Make these changes to your plotting sheet:

A3: 0
C2:

```
=IF(0.707*O.XLM!x2s1(C$1,$B2,D)+0.707*
  O.XLM!px2(C$1,$B2,D))^2>0.1,"*"," ")
```

We set the distance between orbitals to zero since we will be combining atomic orbitals that are on the same atom. 0.707* needs to be inserted two places in the current cell C2 and px1 changed to 2s1. This will combine a 2s and a 2p orbital to make an sp hybrid such as we might find on carbon in C_2H_2. After copying cell C2 from C2 through AI22, you should see an sp hybrid orbital as in Worksheet 10.4.

Change C2 again by changing the + sign to a - sign and copying into cell C2 through AI22. This should give you an identical shaped orbital but facing opposite to the first.

	A	B	C	D	E	F	G	H	I	J	K	L	M	N	O	P	Q	R	S	T	U	V	W	X	Y	Z	AA	AB	AC	AD	AE	AF	AG	AH	AI
1			-5	-5	-5	-4	-4	-4	-3	-3	-3	-2	-2	-2	-1	-1	-1	0	0	0	1	1	1	2	2	2	3	3	3	4	4	4	5	5	5
2		-3																																	
3	0	-3																																	
4		-3																																	
5		-2											*	*	*	*	*																		
6		-2										*	*	*	*	*	*	*																	
7		-2									*	*	*	*	*	*	*																		
8		-1								*	*	*	*	*	*	*	*																		
9		-1								*	*	*	*	*	*	*																			
10		-1							*	*	*	*	*	*	*	*			*	*	*														
11		0							*	*	*	*	*	*	*	*		*	*	*	*	*													
12		0							*	*	*	*	*	*	*	*		*	*	*	*	*													
13		0							*	*	*	*	*	*	*	*		*	*	*	*	*													
14		1							*	*	*	*	*	*	*	*			*	*	*														
15		1								*	*	*	*	*	*	*																			
16		1								*	*	*	*	*	*	*	*																		
17		2									*	*	*	*	*	*	*																		
18		2										*	*	*	*	*	*	*																	
19		2											*	*	*	*	*																		
20		3																																	
21		3																																	
22		3																																	

Worksheet 10.4 An sp Hybrid Orbital

The atomic orbitals have been combined in two separate ways to obtain the two separate sp hybrid orbitals. We added 2s and $2p_x$ orbitals together to get one and we subtracted one from the other to get the second. Notice the larger lobes on the two sp hybrid orbitals are pointing in opposite directions 180° apart. This allows a linear arrangement of bonds if both are used to form molecular orbitals.

To form the three sp^2 hybrids, the 2s, $2p_x$, and $2p_z$ orbitals are combined. Change C2 and copy as before. Notice the $2p_z$ orbital is not used yet. It will be in the last two orbitals. The first 0.707 is changed to 0.577 and the second to 0.816 to make the change.

C2:

```
=IF((0.577*O.XLM!x2s1(C$1,$B2,D)+0.816*
   O.XLM!px2(C$1,$B2,D))^2>0.1,"*"," ")
```

This sp^2 hybrid is similar in shape to the sp hybrids. The second sp^2 hybrid is made by combining the 2s, $2p_x$, and $2p_z$ orbitals as follows.

C2:

```
=IF((0.577*O.XLM!x2s1(C$1,$B2,D)-0.408*O.XLM!px2(C$1,$B2,D)
        -0.707*O.XLM!pz2(C$1,$B2,D))^2>0.1,"*"," ")
```

This can be obtained by editing the previous expression. Two numbers are changed along with the sign on one of them and the $2p_z$ orbital is added. When the sheet is run, this plots an orbital pointing to the lower right at 120° from the first one. Now for the third orbital.

C2:

```
=IF((0.577*O.XLM!x2s1(C$1,$B2,D)-0.408*O.XLM!px2(C$1,$B2,D)
      +0.707*O.XLM!pz2(C$1,$B2,D))^2>0.1,"*"," ")
```

This differs from the previous expression by only one sign change on 0.707. Copy C2 into C2 through AI22. Now only one p orbital is subtracted while the other is added and the resulting hybrid is located 120° from each of the first two, giving a trigonal or triangular planar geometry.

Other types of hybrid orbitals are produced in a similar manner by using different combinations of atomic orbitals, including d orbitals in some cases, to obtain the geometry needed for a particular compound. These can then be used to form molecular orbitals by overlapping them with atomic or hybrid atomic orbitals on other atoms. To see one such molecular orbital between an sp hybrid and a $2p_x$ orbital, make the following changes:

A3: 9
C2:

```
=IF((0.707*O.XLM!x2s1(C$1,$B2,D)-0.707*
O.XLM!px1(C$1,$B2,D)+O.XLM!px2(C$1,$B2,D))^2>0.1,"*"," ")
```

Copy C2 into C2 through AI22. Running the sheet with this change shows part of the two separate orbitals to be overlapped. Try R = 8, 6.5, 5, and 3 Å to see the molecular orbital, and thus the bond, form as the atoms come together. Just as before,there is a buildup of electron density between the two nucleii, and the electrons formerly restricted in the two separate orbitals are now able to move back and forth, being shared by the two atoms as is typical of a covalent bond.

11

Gas Laws

The behavior of gases as pressure, temperature, volume, and the number of moles change is described by equations of state, with the most well known and easiest to use being the ideal gas law. This equation is very useful in many common situations and is thus the most frequently used, but at high pressures or other conditions where the molecules are closer together, it must be replaced by better, although more complicated, equations. The most common of these is the van der Waals equation. This chapter will discuss the properties of gases using these two equations in some detail. We will explore under what conditions each equation can and should be used. (BLB Chap. 10)

The Ideal Gas Law

An ideal gas is one in which the attractive forces between molecules and the finite volume of the molecules can be ignored. These are good assumptions at modest pressures (1 atm) and temperatures (298 K). In fact, experimental measurements under these conditions led to laws about how the different gas variables affected each other. For example, the volume is proportional to the temperature, the volume is proportional to the number of moles of gas (note this is independent of the actual compound), and the volume is inversely proportional to the pressure. In equation form, these statements say that the volume equals a constant value multiplied by the variable in question if proportional and divided by that variable if inversely proportional:

$$V = const \times T$$

$$V = const' \times n$$

$$V = const''/P$$

where T is the absolute temperature, n is the number of moles, and P is the pressure.

If all three of these equations are combined into one, with the overall constant being called R, we get:

$$V = RTn/P$$

or in the form in which it is usually written:

$$PV=nRT$$

where P is the pressure, V is the volume, n is the number of moles, and T is the absolute temperature. The gas constant, R, will have different values depending on the units used for the variables. Two common values are:

R = .0821 L atm/mol K	volume in liters (L), pressure in atmospheres (atm).
R = 8.314 J/mol K	volume in cubic meters (m^3), pressure in Pascals (Pa).

V versus *T*

Let's set up a spreadsheet to study the behavior of the ideal gas volume and how it is affected by temperature change. Worksheet 11.1 shows the start of such a sheet with the number of moles and pressure entered in cells A7 and B3. The formula for solving for volume is entered in cell E7 with the value for temperature being taken from the cell directly above in row 6.

$V = nRT/P$ becomes =$A7*.0821*E$6/$B$3

Notice which row, column, and cell addresses are absolute. These choices will be necessary when the formula is copied into other cells. The volume is calculated to be 24.63 liters.

Now if the number of moles is extended from cell A7 to A10 and the formula in E7 is copied from E7 through H10, volumes for a number of temperatures for different numbers of moles are calculated. This is shown in Worksheet 11.2.

Next select the shaded area shown in Worksheet 11.2 and make a chart by selecting File, New, Chart from the menus to see a plot of this data. You can add headings and legends if you wish as shown in Chart 11.2. Note that the lines are all straight and when extrapolated back to lower temperature, they all intersect at 0 K. This corresponds to the lowest temperature possible (absolute zero) and was first determined by plots of this type.

	A	B	C	D	E	F	G	H
1	**Ideal Gas -- Volume versus Temperature**							
2								
3	Pressure =	1	atm					
4								
5	moles				**Temperature (K)**			
6		0	100	200	300	400	500	600
7	1				24.63			

Worksheet 11.1 Spreadsheet for Calculating the Volume of an Ideal Gas as a Function of Temperature

	A	B	C	D	E	F	G	H
1	**Ideal Gas -- Volume versus Temperature**							
2								
3	Pressure =	1	atm					
4								
5	moles				**Temperature (K)**			
6		0	100	200	300	400	500	600
7	1				24.63	32.84	41.05	49.26
8	2				49.26	65.68	82.1	98.52
9	3				73.89	98.52	123.15	147.78
10	4				98.52	131.36	164.2	197.04

Worksheet 11.2 Spreadsheet for Showing Temperature Dependence of Volume for an Ideal Gas

1. Next select Window, Arrange All so that both your spreadsheet and chart are displayed at the same time. Now change the pressure value on the spreadsheet and see how the volume values are affected. Notice the change on the plots on the chart. (The chart may not appear to change much, but look carefully at the numbers on the volume axis.) Could you predict how the volume would change when the pressure was changed from the ideal gas law? This will be explored in detail in the next section. You can also change the number of moles and observe the changes. Could you predict these changes? Last of all. pick a different set of evenly spaced temperatures such as 0, 500, 1000, 1500, 2000, etc., and observe the effect. Do these changes match what you expected? Can you find any values of these variables where the plots do not extrapolate to 0 K?

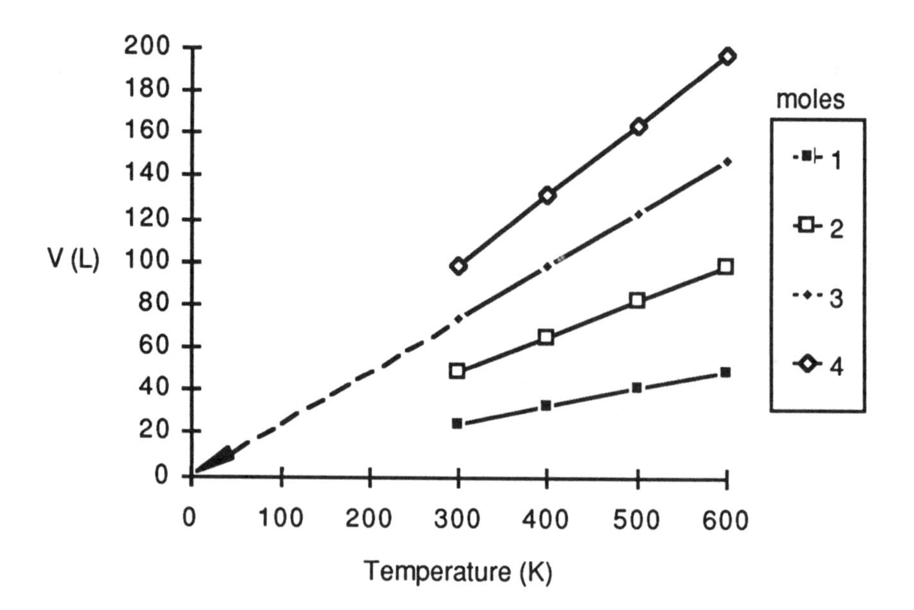

Chart 11.2 Plots of Volume Versus Temperature for Several Amounts of Ideal Gas at 1 atm Pressure

V versus *P*

Now change the spreadsheet in Worksheet 11.2 so that the volume is calculated at different pressures and temperatures. The locations to change are shown shaded in Worksheet 11.3. Enter the formula, V = nRT/P, in cell E7 as:

=$B$3*.0821*$A7/E$6

2. Then copy this formula from C7 through H10. The results should look like Worksheet 11.3. Now select cells A6 through H10 and make a chart. Your plots should look like those in Chart 11.3. Notice the reciprocal nature of the volume and pressure. As one goes up the other goes down. Select Window, Arrange All from the menu so you can see both the chart and the sheet. Change the number of moles and observe the changes. Change the temperatures and see what happens. Then put in some new evenly spaced pressures (such as 0, 0.1, 0.2, etc., or 0, 10, 20, etc.) and see what happens. Did all of these changes happen as you expected?

	A	B	C	D	E	F	G	H
1	Ideal Gas -- Volume versus Pressure							
2								
3	moles =	1						
4								
5	Temp (K)				Pressure (atm)			
6		0	1	2	3	4	5	6
7	300		24.63	12.315	8.21	6.1575	4.926	4.105
8	400		32.84	16.42	10.9467	8.21	6.568	5.47333
9	500		41.05	20.525	13.6833	10.2625	8.21	6.84167
10	600		49.26	24.63	16.42	12.315	9.852	8.21

Worksheet 11.3 Sheet for Calculating Volume Versus Pressure for Different Temperatures and Amounts of Ideal Gas

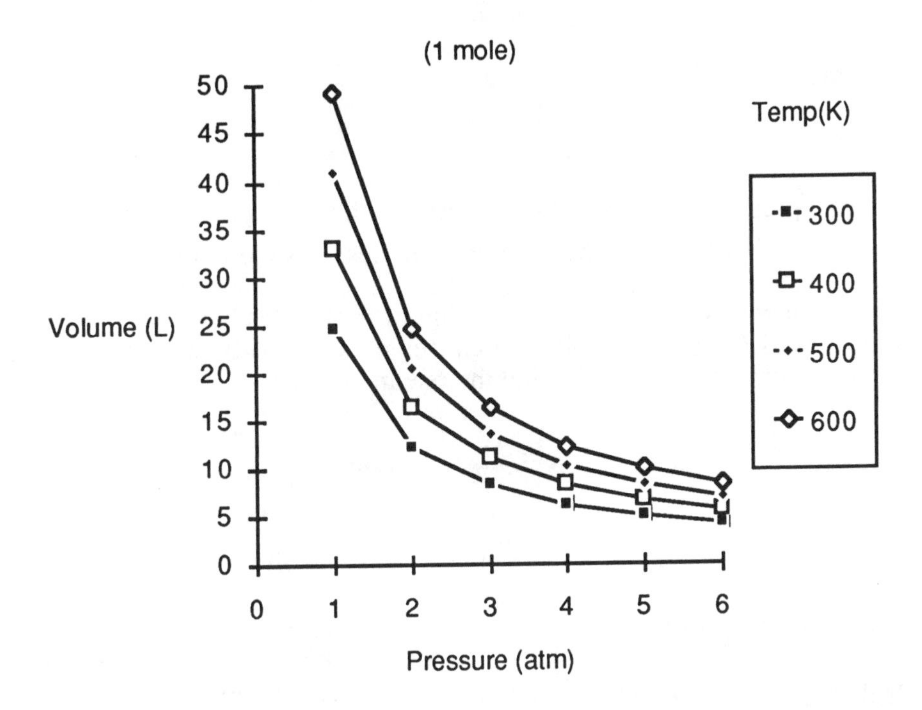

Chart 11.3 Plots of Volume Versus Pressure for an Ideal Gas

3. The sheet can be changed again to show the effect of changing the number of moles on the volume. Change moles to pressure and pressure to moles. The formula in cell E7 that we have been modifying in the previous examples should be changed to reflect these

modifications. A chart can then be made and should show the linear relationship between volume and moles.

Van der Waals Gases

At high pressure and/or low temperature, where the volume gets smaller and the molecules closer together, the ideal gas law breaks down because the assumptions behind it (no attractive forces between molecules and no significant molecular volume) are no longer true. An equation of state that takes these factors into account then needs to be used. The most common one is the van der Waals equation. Its popularity may be due to the direct way these complicating factors are taken into account. The equation is:

$$(P + an^2/V^2)\ (V-nb) = nRT$$

Compare this to the ideal gas law:

$$(P_{ideal})(V_{ideal}) = nRT$$

The actual pressure is P. This is corrected to the ideal pressure, P_{ideal}, by the term an^2/V^2 where n is the number of moles, V is the volume, and a is a measure of the magnitude of the attraction between the molecules. This essentially takes into account the slowing of molecules headed toward the container wall (where the collision causes pressure to be exerted) by the attraction of other molecules behind the one in question.

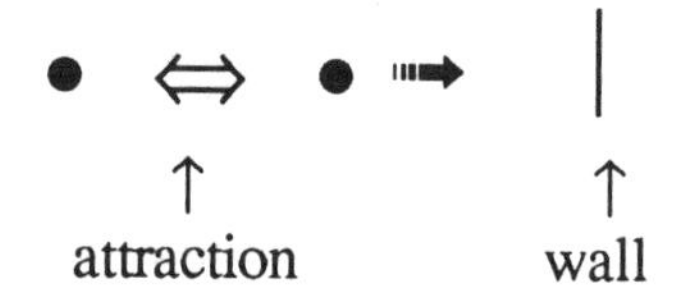

The actual volume, V, which includes the empty space between molecules plus the volume of the molecules themselves, is corrected to V_{ideal} ,where the molecules have no volume themselves, by the term nb where n is the number of moles and b approximates the actual volume of the molecules themselves per mole.

The forces and volumes cannot be calculated theoretically and so are obtained experimentally for each gas by actually measuring all the variables under different conditions. This of course leads to different values for a and b for each gas and so in effect we have a different equation for each gas. This is not as convenient as using one equation for all gases as we did for ideal gases, but under certain conditions, which will be explored later, it is necessary to live with this complication. Table 11.1 lists some typical values for the van der Waals constants in order of increasing attractive force.

Table 11.1 Selected van der Waals Gas Constants

Gas	*a* (atm L^2/mol^2)	*b* (L/mol)
Ideal	0.0	0.0
He	0.0341	0.0237
H_2	0.244	0.0266
N_2	1.39	0.0391
CH_4	2.25	0.0428
CO_2	3.59	0.0427
Xe	4.19	0.051
H_2O	5.45	0.0305
C_2H_6	5.48	0.0638
Cl_2	6.49	0.0562
C_3H_8	8.64	0.0845
CCl_4	20.4	0.1383

The different *a* values correspond roughly to the differences in forces between molecules as measured by differences in the normal boiling points of the materials, with higher *a* values matching up with higher boiling points. For example, the *a* values and boiling points for CH_4 are 2.25 and -161°C, for C_2H_6 are 5.48 and -104°C, and for C_3H_6 are 8.64 and -42°C. There are notable exceptions to this, for example H_2O with values of 5.45 and +100°C. This can be explained by noting that in the gas phase the hydrogen bonding, that is strong in the liquid and solid, will have little effect, and so the *a* value for water is similar to what it would have if only its permanent dipole and the usual London dispersion forces were considered. In fact, extrapolating the boiling points of H_2S and H_2Se (-62°C and -42°C), would give water a boiling point of about -80°C. This is about the value water would have if the hydrogen bonding didn't exist and thus, water would be very similar to C_2H_6.

The *b* values are in liters per mole and correspond roughly to the volume of the material in the liquid or solid state where presumably there is little empty space between the molecules. For example, 1 mol of H_2O has a mass of about 18 g and since the density of water is about 1 g/cm^3, the volume of one mole of water is 18 cm^3 or about .018 L. This compares to the van der Waals *b* value for H_2O of .030 L/mol. This is typical of the agreement between actual volumes and the b constants for various compounds.

Worksheet 11.4 shows the start of a spreadsheet to calculate the volume of a van der Waals gas. The number of moles are input in A3, the pressure in C3, and a series of temperatures in D5 through J5. The van der Waals constants, a and b, are in cells A7 and B7. Cell F6 contains the formula for the volume of an ideal gas for comparison (=$A$3*.0821*F$5/$C$3). Cell F7 contains a formula for the volume of the van der Waals gas. Since the van der Waals equation becomes a cubic equation in volume when it is rearranged, the iterative feature of Excel is used by writing the equation as:

$$V = \frac{nRT}{P + \frac{an^2}{V^2}} + nb$$

Since the iteration starts at $V = 0$, a divide by zero error will occur, so the equation is modified by multiplying the numerator and denominator of the first term by V^2 giving:

$$V = \frac{V^2 nRT}{V^2 P + an^2} + nb$$

This is entered into cell F7 as:

```
=IF($E$3="set",F$6,F7^2*$A$3*.0821*F$5/(F7^2*$C$3
        +($A7*$A$3^2))+$A$3*$B7)
```

Be sure to then select Options, Calculation..., Iteration so the cell can calculate a value with itself as a variable in the formula. The IF allows you to set the initial value of cell F7 to that of the volume of an ideal gas by typing "set" into cell E3. Then changing cell E3 to anything else, such as "it", allows the iteration to proceed. This initialization is necessary because in some cases, as variables such as temperature are changed, the wrong solution for volume is sometimes found. This can happen since the equation is equivalent to a cubic polynomial that can have three different roots, with only one of them being the correct chemical value. Which root is found in the iteration depends on the starting value of V, so we must set it each time we change variables. This setting of the initial value will be used in all of the remaining worksheets in this chapter.

Worksheet 11.5 shows two more gases added with the formula in cell F7 copied from F7 through J9. The ideal gas formula in F6 is copied from G6 through J6. At this pressure (1 atm), the values are not much different from the ideal gas values. Select cells C5 through J9 and choose File, New,

Chart to make a graph of this data. The result should look like Chart 11.5A. Now select Window, Arrange All so that both the sheet and chart are visible. Next increase the pressure to 10 atm, type "set" in E3, then type "it" in E3, and note the changes on the chart. Chart 11.5B shows the result of this change. Note that the gases are starting to deviate more from ideality. Then change the pressure back to 1 atm and change the temperatures to 0, 25, 50, 75, etc., set initial values again, then set iteration, and note the results in the chart (see Chart 11.5C). Again, the atoms are getting closer together and the gases deviate more from ideality. Also note that those gases with the larger *a* and *b* values deviate more. Note the point at 50 K for chlorine in this last chart. It has dropped essentially to zero.

	A	B	C	D	E	F	G	H	I
1	Volume versus Temperature for van der Waals Gases								
2									
3	1	moles	1	atm	it	<- set initial value or iterate			
4						Temperature (K)			
5	a	b		0	100	200	300	400	500
6	–	–	ideal			16.42			
7	1.39	0.0391	nitrogen			16.3744			

Worksheet 11.4 Spreadsheet for Calculating the Volume of a van der Waals Gas as a Function of Temperature

	A	B	C	D	E
1	Volume versus Temperature for van der W				
2					
3	1	moles	1	atm	it
4					
5	a	b		0	100
6	–	–	ideal		
7	1.39	0.0391	nitrogen		
8	3.59	0.0427	carbon dioxide		
9	6.49	0.0562	chlorine		

	F	G	H	I	J
1	aals Gases				
2					
3	<- set initial value or iterate				
4	Temperature (K)				
5	200	300	400	500	600
6	16.42	24.63	32.84	41.05	49.26
7	16.3744	24.6127	32.8368	41.0553	49.2709
8	16.2423	24.5266	32.7733	41.0052	49.2298
9	16.0738	24.4211	32.6981	40.9479	49.1844

Worksheet 11.5 Spreadsheet Showing Temperature Dependence of Several van der Waals Gases as a Function of Temperautre

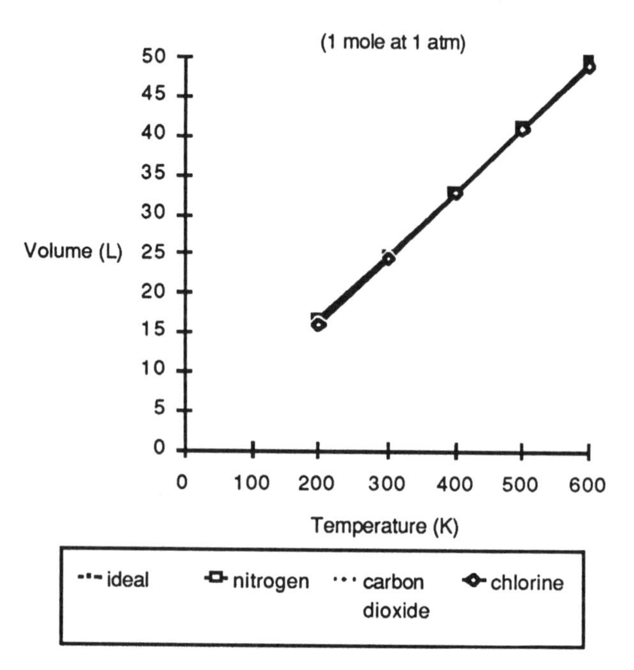

Chart 11.5A Plots of Volume Versus Temperature for Several van der Waals Gases at 1 atm

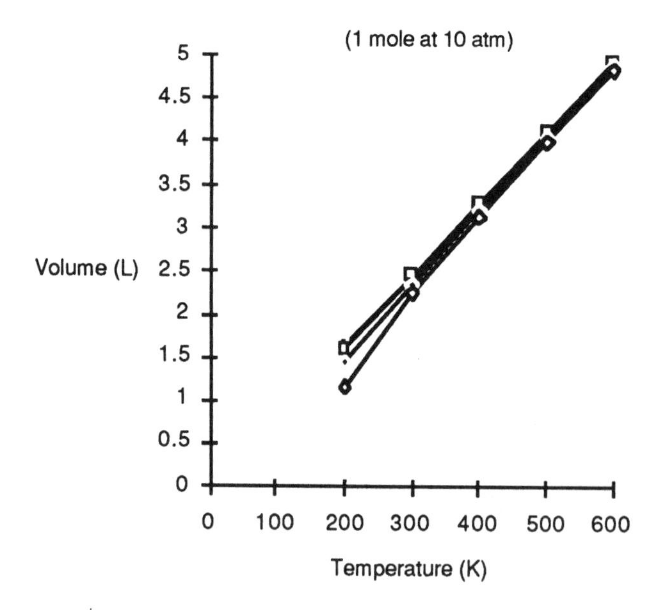

Chart 11.5B Plots of Volume Versus Temperature for Several van der Waals Gases at 10 atm

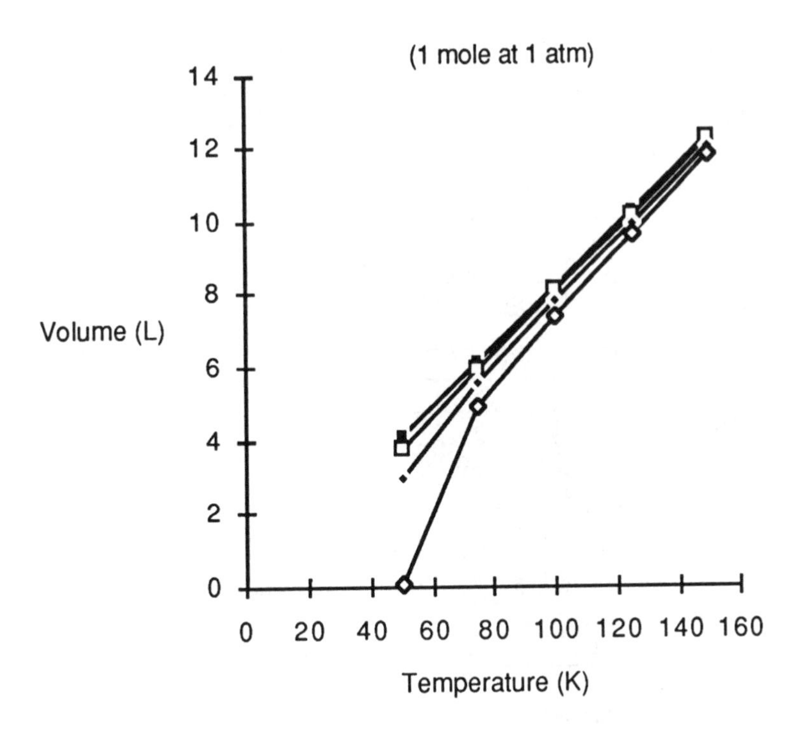

Chart 11.5C Plots of Several van der Waals Gases at 1 atm and Lower Temperatures

Of course the gases will liquify if cooled enough and the van der Waals equation crudely reflects this, although it does not describe liquid behavior and cannot be used to predict the boiling point.

Worksheet 11.6 shows the sheet in Worksheet 11.5 modified to calculate volume for several pressures all at one temperature. The temperature is entered in cell C3. The formula in E6 for an ideal gas is changed to:

```
=$A$3*.0821*$C$3/E$5
```

and copied from E6 to J6. The formula in cell F7 is changed to:

```
=IF($F$3="set",f$6,F7^2*$A$3*.0821*$C$3/(F7^2*F$5
        +($A7*$A$3^2))+$A$3*$B7)
```

Again the IF allows setting the initial values before the iteration by typing "set", followed by "it" in cell F3. This is copied from cell E7 through J9.

	A	B	C	D	E
1	Volume versus Pressure for van der Waals Ga				
2					
3	1	moles	300	degrees K	
4					
5	a	b		0	1
6	--	--	ideal		24.63
7	1.39	0.0391	nitrogen		24.61271
8	3.59	0.0427	carbon dioxide		24.52658
9	6.49	0.0562	chlorine		24.42106

	F	G	H	I	J
1	ses				
2					
3	it	<-set initial value or iterate			
4	Pressure (atm)				
5	2	3	4	5	6
6	12.315	8.21	6.1575	4.926	4.105
7	12.29777	8.192816	6.140368	4.908919	4.087972
8	12.21122	8.105848	6.052972	4.82109	3.999703
9	12.10436	7.997603	5.943285	4.709902	3.886949

Worksheet 11.6 Spreadsheet for Calculating Volume of Several van der Waals Gases as a Function of Pressure

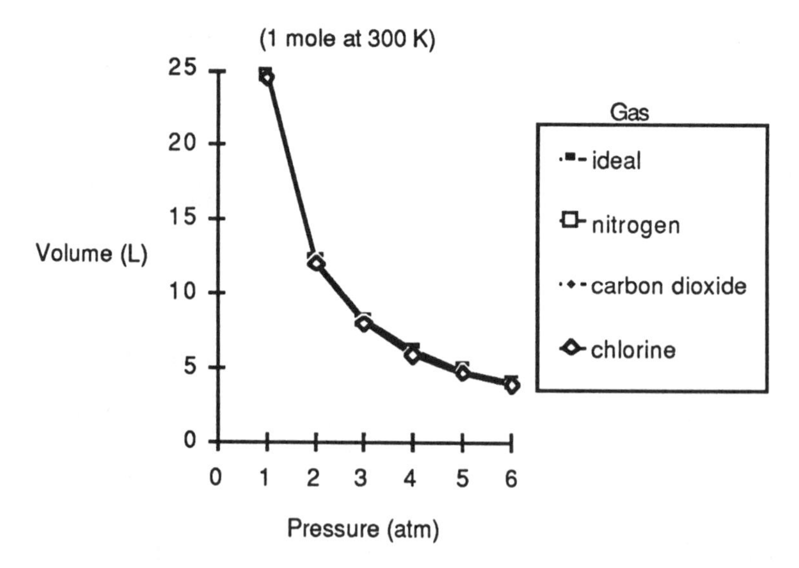

Chart 11.6A Plots of Volume Versus Pressure for Several van der Waals Gases at 300 K

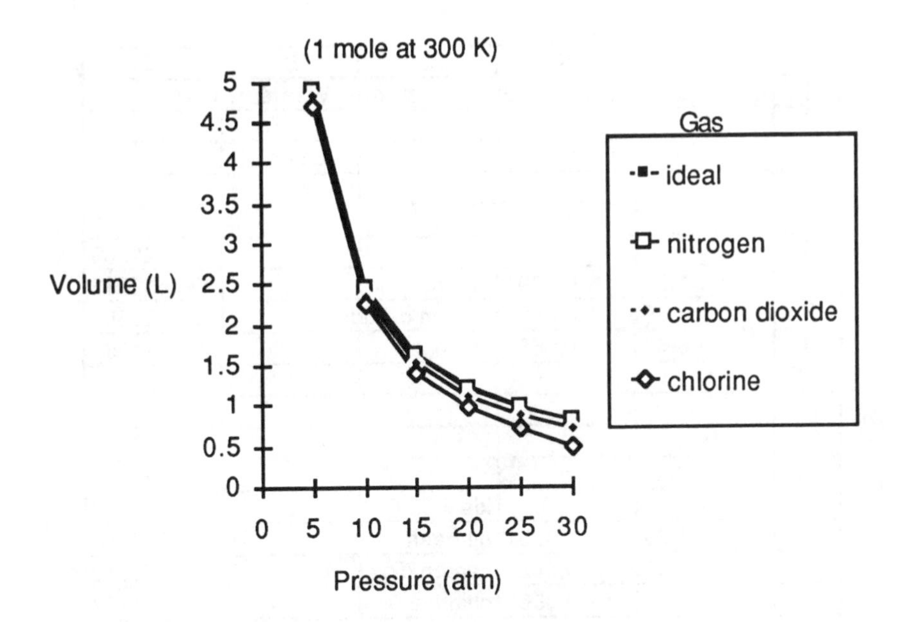

Chart 11.6B Plots of Volume Versus Pressure for Higher Pressures

Charts 11.6A and 11.6B show plots from this sheet for different pressure ranges, again showing the deviation from ideal as the pressure goes up.

The compressibility factor, $z = PV/nRT$, is very sensitive to deviations from ideal behavior. This quantity equals one for an ideal gas regardless of the pressure. The Worksheet 11.6 sheet is extended in Worksheet 11.7. Cells A5 through C9 are copied down to A13 through C17. The pressures in D5 to J5 are copied to D13 to J13 and 1's are placed in D14 to J14 and D14 to D17. The formula for *z* is placed in cell E15 and copied from E15 through J17:

= E7*E$5/($A$3*.0821*$C$3)

Select C13 through J17 and make a chart as in Chart 11.7A. You can see that chlorine and carbon dioxide deviate several percent from ideal whereas nitrogen is still very close to ideal. Now change the pressure range to 0, 10, 20, etc., initialize the starting values, start iteration, and make another chart. Chart 11.7B shows the deviations are even greater at these higher pressures. Make the pressure range 0, 50, 100, etc., to see even greater deviations as in Chart 11.7C. The deviations below the ideal line are due to attractive effects where the pressure is lower than ideal. The gases at higher pressures are affected increasingly more by the volume effects, and with the real volume being larger than the ideal volume, the curve goes above the ideal line.

	A	B	C	D	E
1	**Compressibility factor -- van der Waals Gases**				
2					
3	1	moles	300	degrees K	
4					
5	a	b		0	1
6	--	--	ideal		24.63
7	1.39	0.0391	nitrogen		24.612715
8	3.59	0.0427	carbon dioxide		24.526583
9	6.49	0.0562	chlorine		24.421057
10					
11					
12					
13	a	b		0	1
14	--	--	ideal	1	1
15	1.39	0.0391	nitrogen	1	0.9992982
16	3.59	0.0427	carbon dioxi	1	0.9958012
17	6.49	0.0562	chlorine	1	0.9915167

	F	G	H	I	J
1					
2					
3	it	**set initial value or iterate**			
4	**Pressure (atm)**				
5	2	3	4	5	6
6	12.315	8.21	6.1575	4.926	4.105
7	12.297765	8.1928163	6.1403677	4.9089195	4.0879717
8	12.211218	8.1058476	6.0529717	4.8210901	3.9997026
9	12.104359	7.9976031	5.9432843	4.7098986	3.8869414
10					
11					
12	**Pressure (atm)**				
13	2	3	4	5	6
14	1	1	1	1	1
15	0.9986005	0.997907	0.9972176	0.9965326	0.9958518
16	0.9915727	0.987314	0.9830242	0.9787028	0.974349
17	0.9828956	0.9741295	0.9652106	0.9561305	0.9468797

Worksheet 11.7 Compressibility Factor Calculations for Several van der Waals Gases

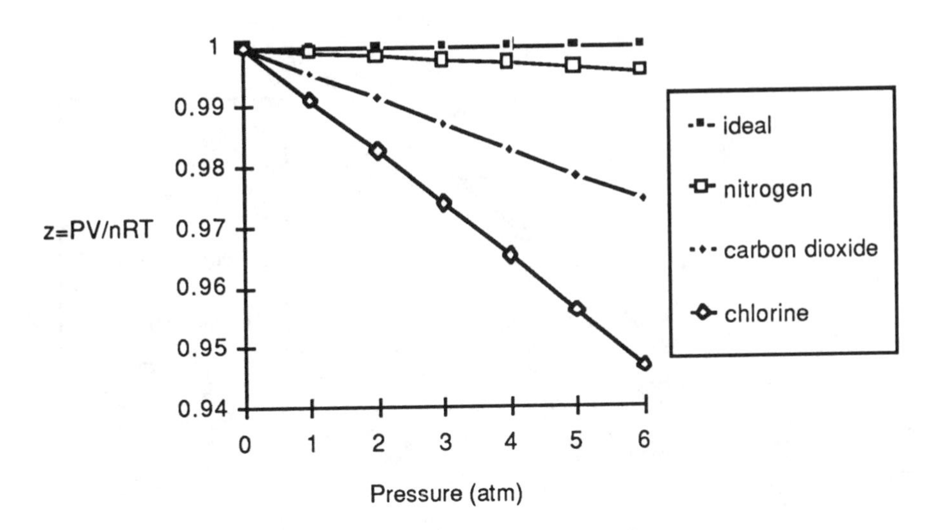

Chart 11.7A Plots of the Compressibility Factor Versus Pressure for Several van der Waals Gases

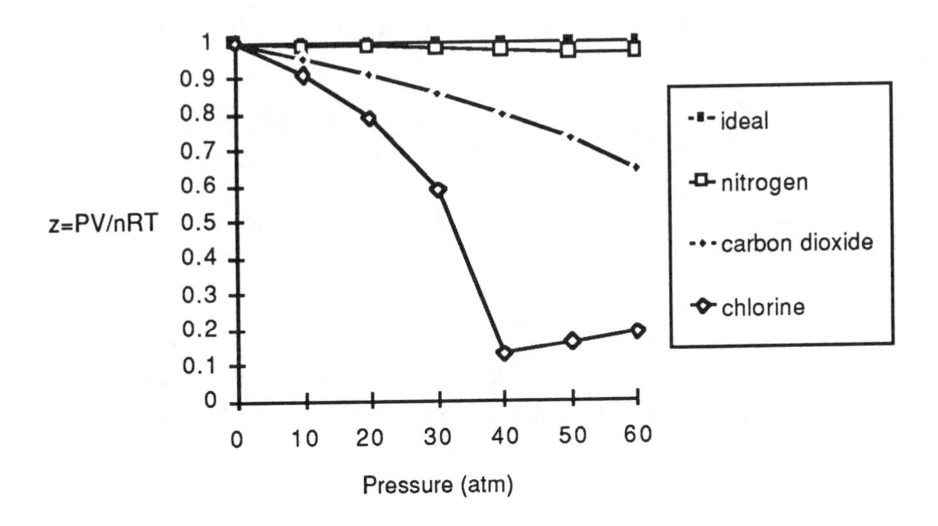

Chart 11.7B Compressibility Plots at Higher Pressures Showing More Deviation

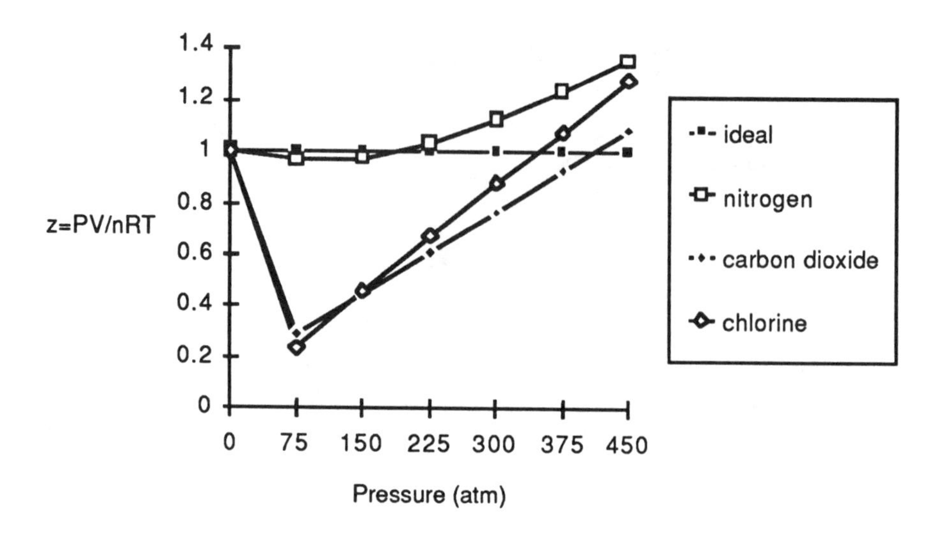

Chart 11.7C Plots of Compressibility Factors for Several van der Waals Gases at Very High Pressures

Worksheet 11.8 shows the last sheet modified to calculate *z* for one gas (nitrogen in this case) for several temperatures. Cell C7 contains the formula:

=IF($E$3="set",$A$4*.0821*$A7/C$6,($A$4*.0821*$A7*C7^2
/(C7^2*C$6+($C$4*$A$4^2))+$A$4*$E$4))

This is copied into cells C7 through H9. Cell C14 contains:

=C$6*C7/($A$4*.0821*$A7)

and is copied into cells C14 through H16. Chart 11.8 is a chart made from cells A12 through H16 and plotted as a scatter chart. This sheet and chart show that the higher the temperature gets, the more closely the gas obeys the ideal gas law.

	A	B	C	D
1	**Compressibility Factor - Nitrogen versus i**			
2				
3				
4	1	moles	1.39	a
5	Temp(K)			
6		0	100	200
7	200		0.1279255	0.0767853
8	400		0.3304093	0.1720901
9	1000		0.8444009	0.4350609
10				
11				
12		0	100	200
13	ideal	1	1	1
14	200	1	0.7790835	0.9352657
15	400	1	1.0061186	1.0480518
16	1000	1	1.0285029	1.0598317

	E	F	G	H
1	**deal gas**			
2				
3	it	**<- set initial value or iterate**		
4	0.0391	b		
5		**Pressure (atm)**		
6	300	400	500	600
7	0.0653534	0.0599812	0.0567157	0.0544667
8	0.1228558	0.1000403	0.0871942	0.0790225
9	0.2993088	0.2318913	0.1917587	0.1652304
10				
11		**Pressure (atm)**		
12	300	400	500	600
13	1	1	1	1
14	1.1940336	1.4611746	1.7270305	1.9902555
15	1.1223123	1.2185178	1.3275608	1.4437727
16	1.0936984	1.1297992	1.1678359	1.2075302

Worksheet 11.8 Spreadsheet for Calculating z for Nitrogen at Several Different Temperatures

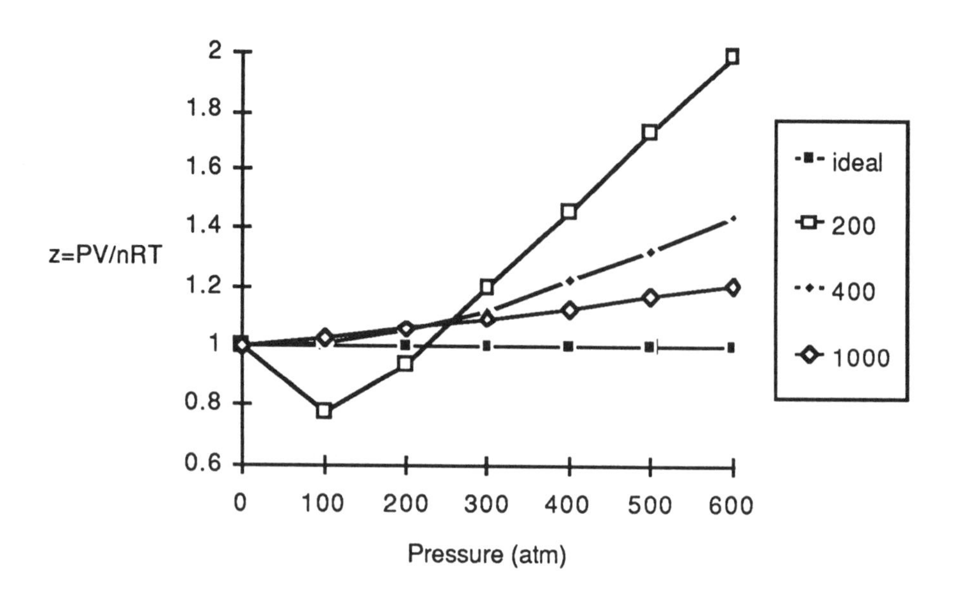

Chart 11.8 Plots of z Versus Pressure for Nitrogen at Several Temperatures

Problems

You should have verified the basic relationships for an ideal gas going through the exercises numbered (1) through (3) in the ideal gas section. The following problems deal with van der Waals gases and their comparison to the ideal gas.

4. Modify the sheet in Worksheet 11.5 to calculate volume for the gases He, H_2O, and CCl_4. Make plots similar to Charts 11.7A through 11.7C. Which of these gases deviates the most from ideal? Which deviates least? At what temperature does CCl_4 try to "liquify"? Repeat these calculations and charts for the series CH_4, C_2H_6, and C_3H_8. Answer the same questions for these three gases. Now consider all of your results here and the examples in the chapter. Which gases are described very well by the ideal gas law at 1 atm and 300 K? Which gases deviate first as the pressure goes up and the temperature goes down?

5. Repeat the above analysis using the sheet in Worksheet 11.6 and the charts in Charts 11.6A through 11.6B. Also try modifying the number

of moles and the temperature. How does increasing the number of moles affect the deviations from ideal. What about lowering and raising the temperature? This should have the same effect as changing temperature in the last problem. Does it? At what pressure do Cl_2, C_3H_8, and CCl_4 try to "liquify" at 300 K?

6. Using the compressibility factor sheet in Worksheet 11.7, calculate *z* for other gases in Table 11.1. Do the same gases show the most deviations from ideal as they did in problems 5 and 6? A percent deviation from ideal can be calculated from this data. For example looking at Worksheet 11.7, *z* = 0.98 for chlorine at 2 atm and 300 K. The deviation can be calculated by the following equation:

$$100 \times |z - 1| = 100 \times |0.98-1| = 2\%$$

where the vertical bars mean take the absolute value (i.e., drop the minus sign). You could add to or modify the sheet in Worksheet 11.7 to do this calculation. Calculate these deviations for several other gases at 0.2 atm, 2 atm, 20 atm, and 200 atm. If we decide that a 1% deviation is all we will allow as a deviation from ideal, at what pressure should we start using the van der Waals equation of state instead of the ideal gas law for each gas in Table 11.1 assuming a 300 K temperature? If we leave the pressure at 1 atm at what temperatures do we need to start using the van der Waals equation of state for the various gases?

7. Modify the sheet in Worksheet 11.8 to calculate *z* at different temperatures for another gas such as Cl_2. Plot this data as in Chart 11.8. How do the changes in the van der Waals constants, *a* and *b*, affect the appearance of this plot? At what pressures do the deviations from ideal switch from negative to positive for the gases in Table 11.1? How is this crossover point affected by temperature? Do all the gases have a temperature and pressure where the deviations are negative?

12

Gas Kinetic-Molecular Theory

The molecules in a gas are in constant motion, colliding with each other and the walls of their container. The collisions with the walls are responsible for the gas having a pressure. The speeds of individual molecules are constantly changing due to these collisions. The speed is also affected by both the mass or molecular weight of the molecules and the temperature of the gas. (BLB Chap. 10)

Even though the speeds of the individual molecules vary, there is an overall distribution of speeds that is maintained. A typical plot of this distribution is shown in Figure 12.1. The details of this plot will vary with different molecules and temperatures, and we will explore some of these effects as we proceed. Note for now that the curve is not symmetrical. It is stretched out to some extent on the higher speed side.

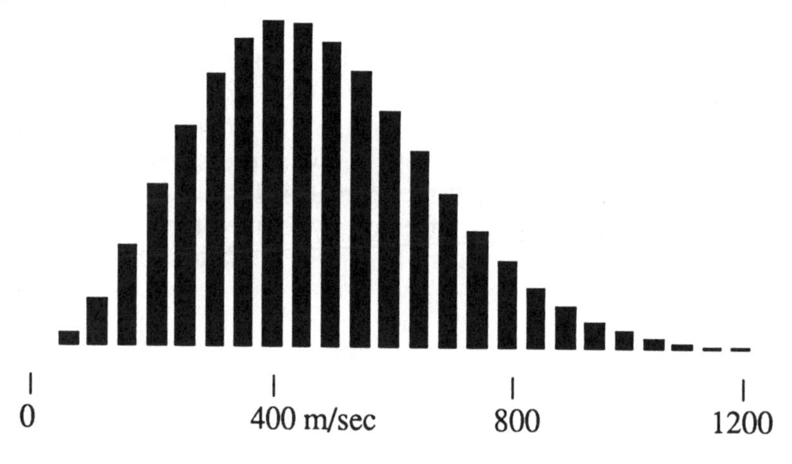

Figure 12.1 Velocity Distribution for Nitrogen at 298 K

The peak of the curve represents the speed more molecules have than any other single speed. This is the most probable speed. There is a simple formula for calculating the most probable speed in terms of the mass of the gas molecules and the temperature of the gas.

$$mps = \sqrt{2RT/M}$$

where R, the gas constant, is 8.314 J/mol K, T is the Kelvin temperature, and M is the molecular weight in kg/mol. Note that the kilograms used in the molecular weight are needed to cancel kg in joules (kg m^2/sec^2). The speed will then be in m/sec.

Let's try a calculation to see what a typical value might be. For nitrogen (28.0 g/mol) at 298 K (room temperature), the most probable speed will be:

$$mps = \sqrt{2(8.314)(298)/.0280} = 421 \text{ m/sec}$$

This then is the most probable speed of the predominant material in the air we breathe. A molecule of N_2 would cover the length of about four football fields in one second if it didn't hit any other molecules. Some molecules of N_2 will be going faster than this speed and some will be going slower. The impact of these molecules with a surface causes the pressure we observe.

Because the distribution curve is stretched toward higher speeds, the average speed will be slightly higher than this most probable speed. It can be calculated with the following formula.

$$avs = \sqrt{8RT/\pi M}$$

or for N_2 at 298 K

$$avs = \sqrt{8(8.314)(298)/3.14159(.0280)} = 475 \text{ m/sec}$$

The velocity distribution curve being stretched out toward higher speeds also makes the median speed (the speed that half of the molecules are above and half are below) not quite equal to the average speed. The formula is

$$mds = 1.538\sqrt{RT/M}$$

or for N_2 at 298 K

$$mds = 1.538\sqrt{8.314(298)/.0280} = 457 \text{ m/sec}$$

The kinetic energy of the molecules can be calculated from their mass and speed.

$$E = (1/2)mv^2$$

where m is the mass in kilograms. The speed, v, needed to calculate the average energy, is the square root of the average of the squares of all the speeds of the molecules. This is called the root mean square speed and can be calculated from:

$$rms = \sqrt{3RT/M}$$

For N_2 at 298 K we get:

$$rms = \sqrt{3(8.314)(298)/.0280} = 515 \text{ m/sec}$$

Notice this is almost identical to the equation for the most probable speed with the only difference being the 3 replacing the 2. If this equation is substituted into the previous equation for energy using the mass for one mole (use the molecular weight in kg/mol), we get:

$$E = (3/2)\, RT$$

for the energy of one mole of gas.

$$E = 3\ (8.314)\ (298)\ /\ 2 = 3716 \text{ joules}$$

for one mole of N_2 at 298 K.

Notice that the energy depends only on the temperature and not on the molecular weight of the gas. Heavier molecules move more slowly than light ones at a given temperature and their energies are thus identical. Increasing the temperature speeds up both the heavy and light molecules and both increase in energy. Let's set up a sheet like the one in Worksheet 12.1 to calculate some of these quantities to further illustrate these points.

We will input the temperature in cell B2 and the molecular weight of the gas in cell B3. Enter the following formula into cell D2 to calculate the root mean square speed:

```
=(3*8.314*$B$2/(0.001*$B$3))^0.5
```

The average speed is calculated in cell F2.

```
=(8*8.314*$B$2/(3.14159*0.001*$B$3))^0.5
```

Next enter the formula for the most probable speed in cell D3.

=(2*8.314*$B$2/(0.001*$B$3))^0.5

Then enter the formula for the median speed in cell F3.

=1.538*(8.314*$B$2/(0.001*$B$3))^0.5

Finally, enter the formula for the energy of one mole of gas in cell D4.

=1.5*8.314*$B$2

	A	B	C	D	E	F
1	Velocity Distribution			(speeds in m/sec)		
2	Temp(K)	298	rms --->	515.2224208	avs --->	474.6837526
3	M.W.	28	mps --->	420.677345	mds --->	457.4993295
4			energy ->	3716.358	J/mol	

Worksheet 12.1 A Sheet for Calculating Several Types of Speeds and the Energy of Gas Molecules

Make sure your sheet calculates the same values as in our previous examples. Then use it to work the following problems.

1. Calculate the most probable speed, the root mean square speed, the average speed, the median speed, and the energy for one mole of each of the following gases at 298 K: H_2, CH_4, CO_2, Xe, and UF_6. What is the effect of changing the molecular weight on these four speeds and the energy?

2. Now calculate these values again at the following temperatures: 5 K, 50 K, 1000 K, and 5000 K. How does cooling a gas change these values? How does heating a gas change these values?

Integrating the Curve

How many molecules are going above a certain speed? How many are going below a certain speed? How many are going between two given speeds? These are questions that can be answered from the curve describing the speed distribution of a gas. If you draw vertical lines at the two speeds on the plot of the speed distribution, as done in Figure 12.2, the fraction of molecules having speeds between those two speeds is just the area under the curve as shown by the white bars.

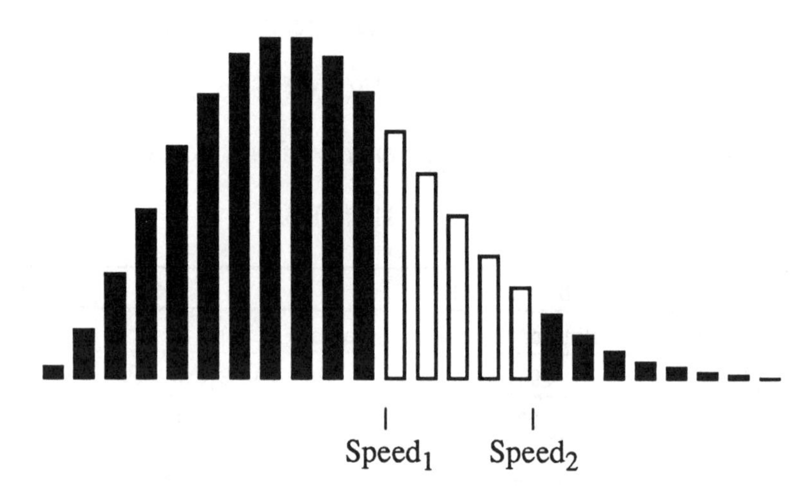

Figure 12.2 Fraction of Molecules Between Two Speeds

The process for calculating this fraction uses calculus and is called integration. Make the following changes to your sheet so it looks like Worksheet 12.2 and you will be able to integrate between any two speeds to calculate the fraction of molecules between them. This sheet also calculates the number of molecules in one mole that are between the two speeds.

We will enter $speed_1$ in cell B4 and $speed_2$ in cell B5. We need to divide this speed range into 100 parts (Δx) to do the calculation, so enter the following formula into cell D5:

=($B$5-$B$4)/50

Next we will put $speed_1$ into cell A7 (=$B$4). Then Δx is added to this speed in cell A8 (=$A7+$D$5). Δx is then added to this cell (A8) in cell A9 and so on until we have 50 speeds between $speed_1$ and $speed_2$. To do this, copy cell A8 into cells A9 through A57. Cell A57 should contain $speed_2$ after the copying is finished.

Now enter the formula describing the velocity distribution into cell B7 (be very careful about where the $'s are placed in each cell reference):

=EXP(-0.001*$B$3*$A7^2/(2*8.314*$B$2))*$A7^2

	A	B	C	D	E	F
1	Velocity Distribution			(speeds in m/sec)		
2	Temp(K)	298	rms --->	515.222421	avs --->	474.683753
3	M.W.	28	mps --->	420.677345	mds --->	457.49933
4	speed1	0	energy ->	3716.358	J/mol	
5	speed2	10000	delta-x	200		
6						
7	0	0	329884.81	1.00000042	<-fraction	
8	200	31907.8389		6.023E+23	<-molecules/mol	
9	400	64784.2718				
10	600	47080.2866				
11	800	17201.9773				
12	1000	3515.07822				
13	1200	421.219902				
14	1400	30.3590836				
15	1600	1.33608334				
16	1800	0.03625554				
17	2000	0.00061066				
18	2200	6.4146E-06				

Worksheet 12.2 A Sheet to Find the Fraction of Molecules Between Two Speeds

Copy this cell (B7) into cells B8 through B57. These values are added together in cell C7:

```
=SUM($B7:$B57)+SUM($B8:$B56)
```

The fraction is calculated in cell D7 with the value normalized so that the total fraction of all the molecules will be equal to 1.

```
=$C$7*4*3.14159*(0.001*$B$3/(2*3.14159*8.314*$B$2))^1.5*$D$5/2
```

The number of molecules between $speed_1$ and $speed_2$ in one mole of molecules is calculated in cell D8.

```
=$D$7*6.023E+23
```

Input N_2 at 298 K. Set $speed_1$ to 0 and $speed_2$ to 10,000. The value of 10,000 for $speed_2$ is large enough in this case to approximate infinity, as there will be essentially no molecules at this velocity and beyond. We will explore this further in the following problems. The result you should get is a fraction very close to 1 (compare your values with Worksheet 12.2 to make sure your sheet is working correctly) since all the molecules will have speeds in this range.

3. What fraction of the molecules are below the most probable speed? Above this speed? Does changing temperature or molecular weight change these values? To answer these questions, input the following sets of data into your sheet. Remember that the most probable speed (*mps*) is calculated and output on your sheet so you can use its value for one of the speeds.

M.W.	Temp (K)	$Speed_1$	$Speed_2$
28.0 (N_2)	298	0	*mps*
28.0	298	*mps*	10,000
28.0	50	0	*mps*
28.0	50	*mps*	10,000
28.0	1000	0	*mps*
28.0	1000	*mps*	10,000
2.01 (H_2)	298	0	*mps*
2.01	298	*mps*	10,000
131.3 (Xe)	298	0	*mps*
131.3	298	*mps*	10,000

Did the two fractions above and below the most probable speed add up to 1 in each case? Notice that the most probable speed changed when the molecular weight or temperature changed. What about the fraction? Are there more molecules above or below the most probable speed?

4. The median speed is the speed that half of the molecules are below and half are above. For N_2 at 298 K, calculate the fraction below the median speed by entering $speed_1 = 0$ and $speed_2$ = median speed (you have this speed calculated on your sheet). Then calculate the fraction above the median speed by using $speed_1$ = median speed and $speed_2$ = 10,000 (infinity). Do these fractions change if the temperature or molecular weight are different? Try some values and see. What are the fractions above and below the average and root mean square speeds? Do these change with temperature or molecular weight?

5. Now let's see how many molecules are above some certain speed. This number can affect how fast reactions take place in the gas phase because collisions between molecules having a certain minimum energy must take place before the reaction can occur. Calculate what fraction of the molecules have speeds above 2000 m/sec for N_2 at 298 K. Use 2000 for $speed_1$ and 10,000 for $speed_2$. Now try the same calculation at 1000 K and see how the fraction increases. Do you think reactions will take place faster at high or low temperatures?

6. Do you think there is a speed above which no molecules are to be found? Run the program for a gas and temperature of your choosing using 5 times the most probable speed as $speed_1$ and infinity (10,000) as $speed_2$. Now try again from 6 times the most probable speed. Then use 7 and 8 times and higher. At what point do you run out of molecules (where does the total number of molecules in one mole become less than 1)?

13

Vapor Pressure

When a liquid is placed in a closed container that is held at a constant temperature, the evaporation of a portion of the liquid results in a vapor pressure above the liquid. This equilibrium vapor pressure of the liquid is a physical property of that liquid at the specified temperature. This vapor pressure is a measure of both the tendency for molecules of the liquid to escape from the liquid phase and for molecules of the vapor phase to condense into the liquid phase at a given temperature. The resulting vapor pressure is, in part, a reflection of the attractive intermolecular forces that are present in the liquid. The vapor pressure of a liquid increases dramatically as the temperature increases, and it decreases as a second substance is added to the liquid at a given temperature. The boiling point of a liquid and the technique of separating components by distillation are consequences of the vapor pressure of the liquid. The elevation of the boiling point of a pure liquid that results from the addition of a second substance is a predictable consequence of the lowering of the vapor pressure. The amount of vapor that results from a particular material is an important environmental and health consideration. (BLB Chap. 11 and 13)

Vapor Pressure as a Function of Temperature

For a specific substance, the partial pressure of the vapor over the liquid is an equilibrium condition that is dependent upon the temperature. The dependence of the vapor pressure on temperature for many substances can be adequately represented as a log function. This equation can be expressed using either the natural (ln) or base ten (log) logarithms. There are several reference books, and in particular the annual *CRC Handbook of Chemistry and Physics*, that tabulate vapor pressure data as two specific constants.

$$\log P = -\frac{0.05223 \cdot a}{T} + b$$

This expression may be viewed as a relationship that is empirical, but it may also be derived from the principles of thermodynamics. The *Clapeyron equation* may be developed from the principle that the Gibbs free energy must be the same for the vapor and the liquid. This equation is a relationship that allows one to calculate the change in pressure for a corresponding change in temperature. For vaporization and sublimation, Clausius showed how the Clapeyron equation may be altered by assuming that the vapor obeys the ideal gas equation and that the molar volume of the liquid is negligible in comparison with the molar volume of the vapor. The results are known as the Clausius-Clapeyron equation.

$$\log P = -\frac{\Delta H_V}{2.303 \cdot R \cdot T} + C$$

In this expression the value of P is in mm of mercury or Torr, R is the gas constant expressed as 8.31452 J/mol-K, C is a constant that depends upon the units of the other parameters, ΔH_V is the enthalpy of vaporization expressed as J/mol, and the liquid temperature, T, is in Kelvin. The natural log form of this equation is similar, but the factor of 2.303 (ln 10) is not necessary when the equation is cast in this form.

Worksheet 13.1 illustrates a spreadsheet that will fit vapor pressure data for two different temperatures to the Clausius-Clapeyron equation. The data used for this worksheet are the familiar vapor pressure data for water. The fit of the data to the equation is not perfect and will change as the selection of the two data points on which to base the logarithmic function is varied. The variation is the result of the fact that the enthalpy of vaporization is actually a slight function of the temperature and not a constant, as the Clausius-Clapeyron equation implies. Extensive data for both inorganic and organic compounds can be found under the heading of vapor pressure in the *CRC Handbook of Chemistry and Physics*. The only data needed for this worksheet are the two temperatures that are entered in column A and the two corresponding vapor pressures that are entered in column B. Column C calculates 1/T and column D calculates the ln P for the two points. These two variables are related in a linear fashion so that $y = m \cdot x + b$ is applicable in the form of $\ln P = m \cdot (1/T) + b$. The slope is calculated from the two-point form of the Clausius-Clapeyron equation as

$$m = (\ln P_2 - \ln P_1) / \left\{\frac{1}{T_2} - \frac{1}{T_1}\right\}$$

The constant b is calculated from the slope and one of the two data points as

$$b = \ln P_1 - m \cdot (1/T_1)$$

	A	B	C	D	E	F	G	H
1	**Vapor Pressure as a Function of Temperature**						delta-H	
2							vapor	b.p.
3	**t(C)**	**P(Torr)**	**1/T**	**ln P**	**Slope**	**Constant**	kJ/mol	T(K)
4	**98**	**707.3**	0.002694	6.5615				
5	**102**	**815.9**	0.002666	6.7043	-4972.04	19.96	41.34	373.2
6								
7		P(Torr)		ln P	P(Torr)exp	Delta-P		
8	0	5.8	0.0037	1.76	4.58	1.20		
9	10	11.0	0.0035	2.40	9.21	1.79		
10	20	20.0	0.0034	3.00	17.54	2.49		
11	30	35.0	0.0033	3.56	31.82	3.22		
12	40	59.2	0.0032	4.08	55.3	3.86		
13	50	96.7	0.0031	4.57	92.5	4.20		
14	60	153.5	0.0030	5.03	149.4	4.06		
15	70	237.1	0.0029	5.47	233.7	3.37		
16	80	357.3	0.0028	5.88	355.1	2.23		
17	90	526.5	0.0028	6.27	525.8	0.75		
18	100	760.0	0.0027	6.63	760	-0.05		

Worksheet 13.1 The Vapor Pressure of Water

The remainder of the formulas used in this spreadsheet are presented in Formula List 13.1. The contents of cell A9 are copied into cells A10 through A18 by using the Fill Down command under Edit on the menu bar. In a similar fashion the contents of cells B8, C8, and D8 are copied into the appropriate cells in rows 9 through 18 using the Fill Down command. If available, experimental data can be entered in column E and the difference between the calculated and experimental values presented in column F. This comparison illustrates that the two parameter, Clasius Clapeyron Equation does not accurately describe the experimental vapor pressure data.

Formula List 13.1 Formulas For Worksheet 13.1

Cell A9	=A8+10	$t_1 + 10$
Cell B8	=EXP(D8)	$e^{(\ln P)}$
Cell C8	=1/(273.15+A8)	$1/(273.15 + t_1)$
Cell D8	=m*C8+b	$m \cdot (1/T_1) + b$
Cell G5	=-0.008314*m	$R \cdot m$
Cell H5	=m/(LN(760)-b)	$m/(\ln P_{atm}) - b$
Cell E5	Defined as m	m
Cell F5	Defined as b	b

A plot of the relationship between the vapor pressure expressed in Torr, *P*, and temperature in degrees Celsius, *t*, is illustrated by Chart 13.1. The dramatic increase in the partial pressure as the temperature increases is characteristic of a logarithmic or exponential function. The rate at which clothing or other objects dry at higher temperatures is illustrated by this graph. When the vapor pressure of a liquid is equal to atmospheric pressure, bubbles of vapor form within the bulk of the liquid and it boils. The temperature at which the vapor pressure of the liquid is equal to the atmospheric pressure is the boiling point of that liquid.

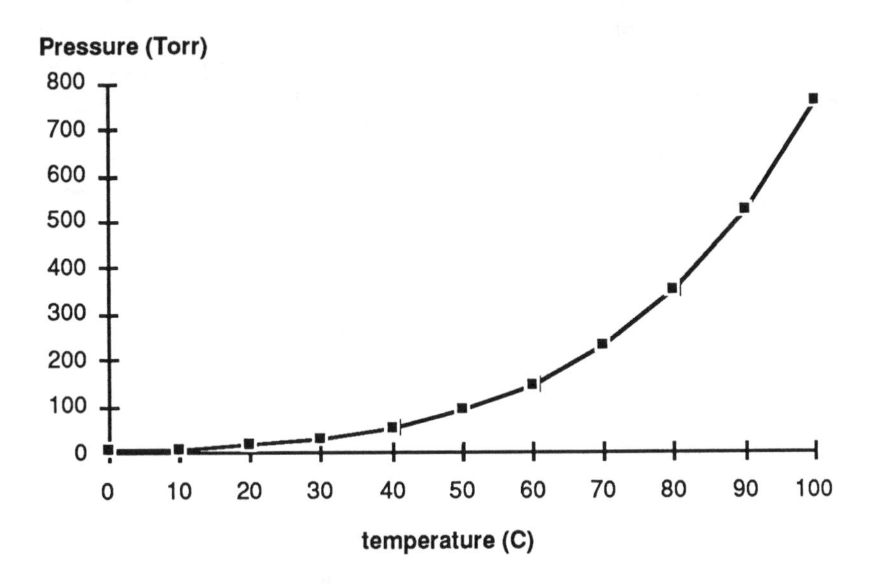

Chart 13.1 Vapor Pressure of Water

The Clausius-Clapeyron equation reveals that a plot of the ln P versus 1/T should be linear, and it is if the temperature interval is carefully chosen. Using this linear relationship, Worksheet 13.1 calculates values for the slope and the constant from two data points, thus assuring that a plot of ln P versus 1/T will be linear. If you attempt to plot the ln P values versus 1/T values directly from this spreadsheet, you will produce a graph that is incorrect because the x-axis scale will be linear and the 1/T intervals decrease as the temperature decreases. A scatter chart option is one method of correcting for the changing size of the 1/T intervals (Appendix C). An alternative method calculates equal size intervals for the values of 1/T in column C. The first and last values of 1/T can be calculated as before, but the intervening values are calculated based on

$$\text{current} = \text{previous} + (\text{last} - \text{first})/10$$

Column A has to be changed so that the highest temperature, 100°C, is the first entry and the lowest, 0°C, is the last. The intervening values for the Celsius temperature are calculated by (1/column C) - 273.15. A plot of this functional relationship is shown as Chart 13.1A for both the experimental values for the vapor pressure of water and for the result calculated from two data points. The open squares designate the linear function as determined by data points at temperatures of 100°C and 34.1°C where the vapor pressures are 760 Torr and 40 Torr. The curve designated by the darkened squares is a plot of the experimental values, and the negative of the slope is equal to the enthalpy of vaporization at that temperature. This plot for water displays an increasingly negative slope as the value of 1/T increases. This means that the enthalpy of vaporization is smaller at the higher temperatures which are closer to the origin. This increase in the enthalpy of vaporization is the result that would be expected if the intermolecular forces are larger at a lower temperature and this is manifested through the clustering of a larger number of water molecules at lower temperatures. This trend dictates that the enthalpy of vaporization for the solid phase (sublimation) is larger than the enthalpy of vaporization for the liquid.

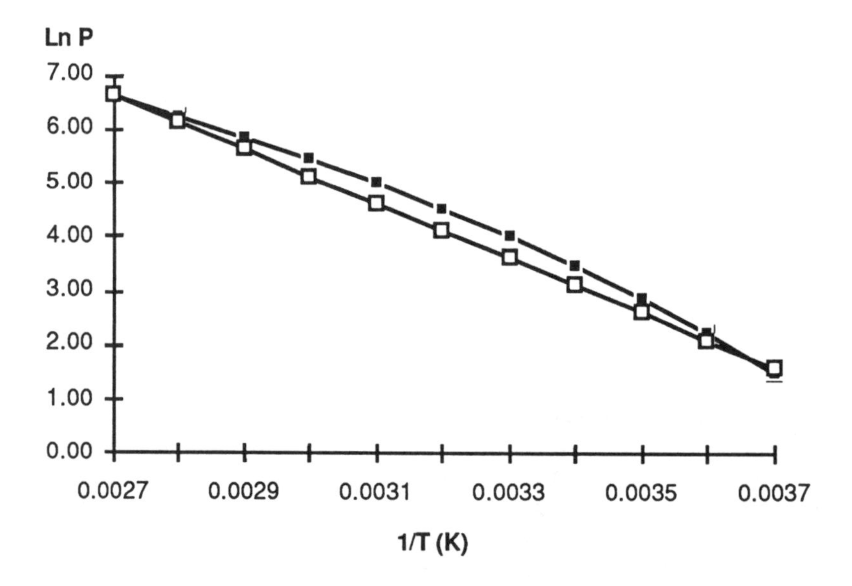

Chart 13.1A Vapor Pressure of Water

Vapor pressure data for several organic and inorganic compounds are given in Table 13.1. More extensive tables can be found in several reference books, such as the *CRC Handbook of Chemistry and Physics* published by The Chemical Rubber Co., Cleveland, OH.

Table 13.1 Vapor Pressure Data

	Temperature °C		
Name	40 mm	400 mm	760 mm
H_2O	34.1	83.0	100.0
C_6H_6	7.6	60.6	80.1
$C_6H_5CH_3$	31.8	89.5	110.6
CCl_4	4.3	57.8	76.7
$CHCl_3$	- 7.1	42.7	61.3
CH_3COCH_3	- 9.4	39.5	56.5
C_6H_{14}	- 2.3	49.6	68.7
CH_3OH	5.0	49.9	64.7
C_2H_5OH	19.0	63.5	78.4
CS_2	- 22.5	28.0	46.5
PCl_3	2.3	56.9	74.2
BBr_3	14.0	70.0	91.7

Vapor Pressure of Binary Solutions

The previous discussion illustrated how the vapor pressure of a liquid increases dramatically as the temperature increases. A lowering of the vapor pressure of a liquid is achieved when a second substance is added to the liquid at a specific temperature. Experimentally, it is observed that the vapor pressure of a solvent over a solution of the solvent and a solute is directly proportional to the mole fraction of the solvent. The slope or proportionality constant is the vapor pressure of the pure solvent. This result is expressed as Raoult's Law.

$$P_1 = X_1 \cdot P_1^\circ$$

An understanding of this phenomena is most apparent when the solute is a nonvolatile material. Chart 13.2 is a plot of the partial pressure or vapor pressure of benzene above a solution of benzene and biphenyl. The data on vapor pressure are measured at 60°C where the vapor pressure of pure benzene is 382 Torr and the vapor pressure of biphenyl is less than 1 Torr. Worksheet 13.2 is used as a basis for graphing the partial pressure of benzene as the vapor pressure of the second component, biphenyl, is of no consequence and may be ignored.

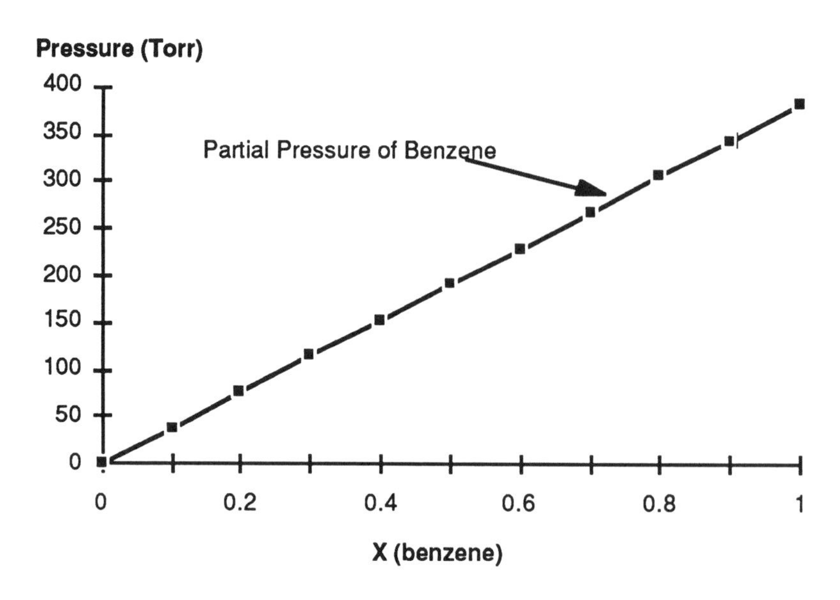

Chart 13.2 Partial Pressure of Benzene/Biphenyl

Raoult's Law is valid when both components are volatile and the partial pressures of both components and the total pressure may be calculated. If the solvent is arbitrarily labeled with a subscript of one, then the solute can be labeled with a subscript two and

$$P_2 = X_2 \cdot P_2^\circ$$

The total pressure is the sum of the partial pressures of the two volatile gases. Worksheet 13.2 calculates the partial pressures of benzene and toluene as a function of the mole fraction of benzene for vapor pressure data observed at 60 °C. Both materials produce a vapor pressure that is linear with respect to the mole fractions of either component. As the partial pressure of the solvent increases in proportion to the mole fraction of the solvent and the vapor pressure of the pure solvent, the partial pressure of the solute decreases in proportion to its mole fraction and the vapor pressure of the pure solute.

The plot of the partial vapor pressure of benzene in a solution containing both benzene and toluene is identical to the plot of the vapor pressure of benzene in a solution of benzene and biphenyl as presented in Chart 13.2. In both cases, the behavior of the vapor was assumed to obey Raoult's Law. A graph or chart of the partial pressures of benzene and toluene, and of the total pressure can be produced by selecting cells A7 through D18 on Worksheet 13.2.

	A	B	C	D
1	**Vapor Pressures of Binary Solutions**			
2				
3	P1 (pure)	382	benzene	60 C
4	P2 (pure)	135	toluene	
5				
6	X(benzene)			
7		P (benzene)	P (toluene)	P (total)
8	0	0	135	135
9	0.1	38	122	160
10	0.2	76	108	184
11	0.3	115	95	209
12	0.4	153	81	234
13	0.5	191	68	259
14	0.6	229	54	283
15	0.7	267	41	308
16	0.8	306	27	333
17	0.9	344	14	357
18	1	382	0	382

Worksheet 13.2 Partial Pressures of Binary Solutions

The formulas for this spreadsheet are presented in Formula List 13.2. The contents of cell A9 are copied into cells A10 through A18 by using the Fill Down command under Edit on the menu bar. In a similar fashion, the contents of cells B8, C8, and D8 are copied into the appropriate cells in rows 9 through 18 using the Fill Down command.

Formula List 13.2 Formulas for Worksheet 13.2

Cell A9	=A8+0.1	$X_1 + 0.1$
Cell B8	=A8*$B$3	$X_1 \cdot P_1^\circ$
Cell C8	=(1-A8)*$B$4	$(1\text{-}X_1) \cdot P_1^\circ$
Cell D8	=B8+C8	$P_1 + P_2$

A graph of the functional relationship between the mole fraction of benzene and the partial pressures of benzene and toluene as well as the total vapor pressure of both above the solution is illustrated by Chart 13.2A. The more volatile component, benzene, has a larger mole fraction present in the vapor phase than is present in the liquid phase for all concentrations except for the condition where only toluene is present ($X_{toluene} = 1$).

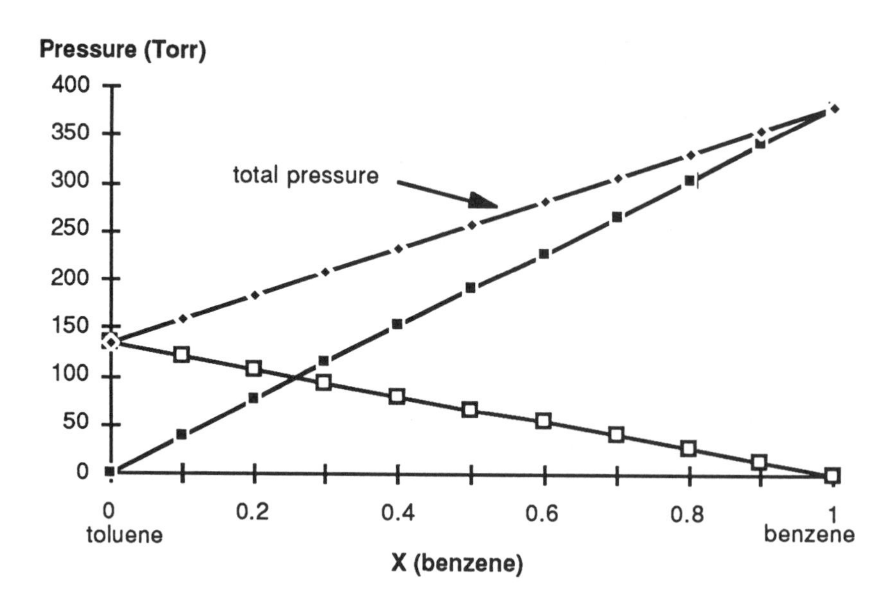

Chart 13.2A Vapor Pressure of Benzene/Toluene

Liquid and Vapor Composition of Binary Solutions

The liquid and vapor composition can be computed for any concentration in a binary solution. As the previous chart illustrates, the vapor phase is richer than the liquid phase in the more volatile component at all concentrations for which the volatile component is present. This is the basis for many separation techniques of which distillation is the most familiar. The mole fraction composition of the vapor phase can be calculated by using a corollary of Dalton's Law of Partial Pressures. The mole fraction of the solvent in the vapor phase can be calculated from

$$X_{1,\text{vap}} = \frac{X_1 \cdot P_1^\circ}{X_1 \cdot P_1^\circ + (1 - X_1) \cdot P_2^\circ}$$

Worksheet 13.3 uses this calculation to illustrate the enrichment of the vapor phase by the more volatile component, benzene. These calculations are shown as column G. In addition, this spreadsheet computes the composition of an imaginary liquid phase, column D, which would produce the vapor composition that is specified by the mole fraction of benzene on the horizontal axis. This curve is know as the vapor composition line. The vapor composition curve illustrates the enrichment by the more volatile component. This liquid-vapor composition graph illustrates the enrichment that can be produced at a specified pressure by a separation technique such as distillation. This calculation requires that the previous equation be solved in terms of the mole fraction of solvent, which results in

$$X_1 = \frac{X_{1\,vap} \cdot P_2^\circ}{(1 - X_{1\,vap}) \cdot P_1^\circ + X_{1\,vap} \cdot P_2^\circ}$$

The result of this calculation of the mole fraction of solvent is then substituted into the following form of Dalton's Law of Partial Pressure.

$$P_{total} = X_1 \cdot P_1^\circ + (1 - X_1) \cdot P_2^\circ$$

This calculation is produced by column C of Worksheet 13.3. Chart 13.3 illustrates a plot of this vapor composition line, which is the vapor pressure of the imaginary liquid phase, on the same graph and mole fraction scale as the total vapor pressure of the original binary liquid. The same scale is used for the mole fraction of benzene in both cases, so the mole fraction values used for the mole fraction in the vapor phase have the same numerical values as the mole fraction in the liquid phase. In both the liquid mole fraction composition and vapor mole fraction composition calculations, the values in column A are used. In some texts this is labeled the mole fraction of benzene composition axis.

	A	B	C	D	E	F	G
1	**Vapor Pressures of Binary Solutions**						
2							
3	P1 (pure)	**382**	benzene	**60 C**			
4	P2 (pure)	**135**	toluene				
5							
6	**Liquid**						**Vapor**
7	X(Bz)						X (Bz)
8		P (total)	P (vapor)		P (Bz)	P (Tol)	
9	0	135	135	0.000	0	135	0.00
10	0.1	160	144	0.038	38	122	0.24
11	0.2	184	155	0.081	76	108	0.41
12	0.3	209	167	0.132	115	95	0.55
13	0.4	234	182	0.191	153	81	0.65
14	0.5	259	199	0.261	191	68	0.74
15	0.6	283	221	0.346	229	54	0.81
16	0.7	308	247	0.452	267	41	0.87
17	0.8	333	280	0.586	306	27	0.92
18	0.9	357	323	0.761	344	14	0.96
19	1	382	382	1.000	382	0	1.00

Worksheet 13.3 Liquid and Vapor Composition

Formula List 13.3 Formulas for Worksheet 13.3

Cell A10	=A9+0.1
Cell B9	=E9+F9
Cell C9	=D9*$B$3+(1-D9)*$B$4
Cell D9	=A9*$B$4/($B$3-A9*$B$3+A9*$B$4)
Cell E9	=A9*$B$3
Cell F9	=(1-A9)*$B$4
Cell G9	=E9/B9

With Worksheet 13.3, column A represents the mole fraction of benzene in the liquid phase. Column E computes the partial pressure of benzene in the vapor phase from the mole fraction of benzene in the liquid phase. In a similar fashion, column F computes the partial pressure of toluene in the vapor phase. The total vapor pressure in the vapor phase is the sum of the partial pressures of benzene and toluene in the vapor phase, and this result is presented in column B. Column B represents the total vapor pressure of both components in the vapor phase as a result of their corresponding mole fractions in the liquid. These computations are identical to those provided by the previous spreadsheet, Worksheet 13.2. Column G is only included in order to emphasize a consequence of unequal vapor pressures. Column G calculates the mole fraction of benzene in the vapor phase. The mole fraction composition of the vapor is richer in benzene for all values except the two extreme values of the mole fraction of either component, when only one component is present. If, at this point, the vapor is condensed to a liquid, this new liquid will be richer in benzene than the original liquid phase, and thus the vapor pressure of the new liquid phase would be higher than the vapor pressure of the original liquid. It is important to realize that the vapor composition curve represents the vapor pressure produced by a liquid phase having a composition that would produce the same mole fraction composition in the vapor phase as that given by the horizontal axis. Column D computes the composition (in terms of the mole fraction of benzene) of a liquid phase that produces the same mole fraction composition in the vapor as the mole fraction composition given in column A. The new mole fraction composition of the solvent that is computed in column D is then used to compute in column C the total vapor pressure a liquid of this composition would produce. The vapor composition curve is shifted relative to the liquid composition curve so that the more volatile component produces a correspondingly higher mole fraction in the vapor phase. A summary of the meaning of the liquid and vapor curves is presented after Chart 13.3. This chart is a plot of the composition of the liquid versus the total vapor pressure produced by this liquid, and the vapor pressure of an imaginary liquid phase having a composition such that the vapor has the same composition as the mole fraction composition on the horizontal axis.

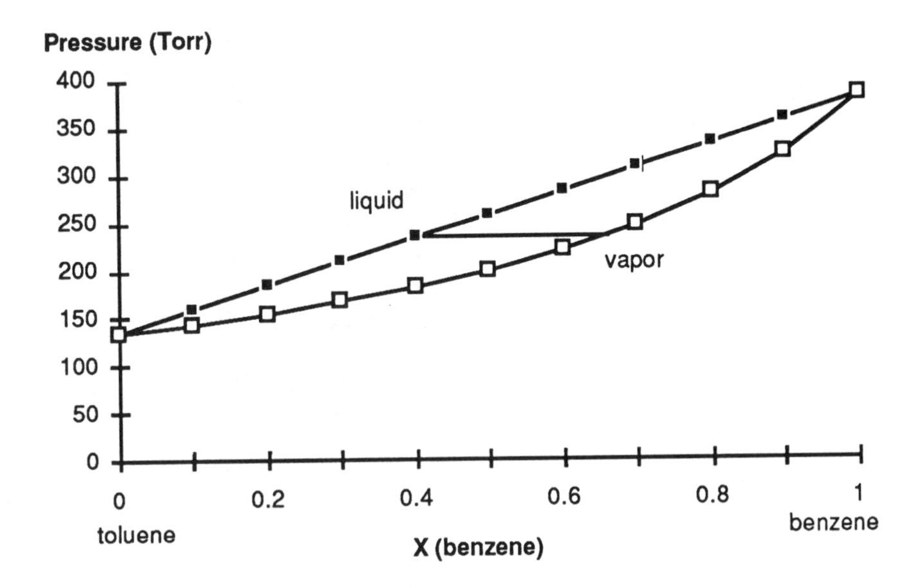

Chart 13.3 Liquid and Vapor Composition

The liquid line is the total pressure of the vapor phase which is in equilibrium with the binary solution having the indicated mole fraction of benzene in the liquid phase. This total pressure line may be considered the composition line for the original liquid at a specified pressure. The vapor line is the total pressure of a vapor produced by an imaginary liquid phase having a composition such that it produces a vapor with the mole fraction composition specified by the horizontal axis. The mole fraction composition of this liquid phase is computed in column D of the spreadsheet. This vapor line uses the same mole fraction of benzene as the liquid line for establishing the vapor composition. The tie line gives the composition of the vapor and the liquid at a specified pressure. The original composition of the liquid is given by the liquid line at a specified pressure. If that pressure is maintained and the vapor physically separated and condensed, it will have the composition given by the vapor line. An alternative interpretation of the vapor curve is that it represents the pressure at which vapors of a specific composition first form a liquid when the pressure is increased.

Boiling Point Elevation

The resulting decrease in the vapor pressure of a solvent when a nonvolatile solute is added means that the solution will have to be raised to a higher temperature in order to promote boiling of the solution. The vapor pressure of the solvent has to be at 760 Torr in both cases in order for boiling to occur. This colligative property is known as boiling point elevation. The spreadsheet that calculated the vapor pressure at temperatures from 0°C to 100°C for vapor pressure data at any two temperatures and pressure can be altered to demonstrate this phenomena. A new column that calculates the reduced vapor pressure as based on Raoult's Law can be added and the results plotted. Chart 13.4 shows the effect of adding a nonvolatile substance to water. The mole fraction used is 0.8 for the solvent, water. The curve for pure water is the upper curve, and the lower curve represents the vapor pressure of water over a solution that has a 0.2 mole fraction of nonvolatile component. Worksheet 13.4 is not shown, you should design a spreadsheet that will compute the vapor pressure of a solvent both with and without a nonvolatile solute present.

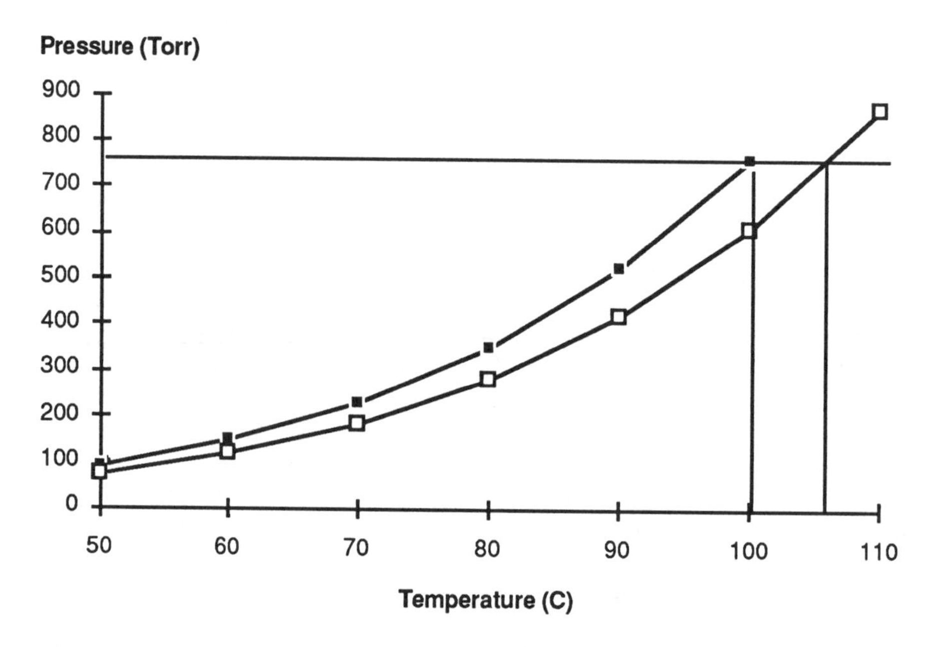

Chart 13.4 Boiling Point Elevation of Water

The addition of a nonvolatile solute to a solvent only changes the constant in the Clausius-Clapeyron equation; the slope is unchanged. In the section on Vapor Pressure as a Function of Temperature, the two-point form of the Clausius-Clapeyron equation was used to calculate the slope from vapor pressure data. The substitution of Raoult's Law in place of the vapor pressure of the pure solvent does not alter the slope, it only changes the intercept. The term representing the mole fraction of the solute disappears when the difference is computed in the slope determination.

$$\ln P^{\circ}_{i+1} - \ln P^{\circ}_{i} = \ln X_1 \cdot P^{\circ}_{i+1} - \ln X_1 \cdot P^{\circ}_{i}$$

The constant is the only quantity that changes when a nonvolatile substance is added to a solvent. The new constant can be evaluated from the vapor pressure data and the resulting vapor pressure of a one molal solution. The boiling point of both the pure solvent and of a one molal solution can be computed. The difference between the boiling point of the pure solvent and that of a one molal solution represents the molal boiling-point-elevation constant. The value of this constant depends on the particular solvent that is selected. Worksheet 13.5 shows the calculation of the molal boiling-point-elevation constant for water where the accepted value is 0.512 °C/molal. Table 13.5 presents additional formulas and the designation of the defined cells. The design of this spreadsheet is very similar to Worksheet 13.1.

	A	B	C	D	E	F	G	H	I
1	Vapor Pressure as a Function of Temperatu						delta-H		
2							vapor	b.p.	b.p.
3	t(C)	P(Torr)	1/T	ln P	Slope	Constant	kJ/mol	T(K)	t(C)
4	83	400	0.00281	5.9915					
5	100	760	0.00268	6.6333	-5017.7	20.080	41.720	373.2	100.0
6									
7		P(Torr)		ln P		mw solvent			
8	100	760.0	0.0027	6.63		18.02			
9	100	746.5	0.0027	6.62	<- 1 molal				
10			20.062	new	constant				
11			373.6	b.p.	T(K)				
12			100.5	b.p.	t(C)				
13			**0.50**	**<- Kb**	C/molal				

Worksheet 13.5 Boiling Point Elevation

	A	B	C	D	E	F
4	**83**	**400**	=1/(273.15+A4)	=LN(B4)		
5	**100**	**760**	=1/(273.15+A5)	=LN(B5)	=(D5-D4)/(C5-C4)=D4-m*C4	
6						
7		P(Torr)		ln P		mw solvent
8	=I5	=EXP(D8)	=1/(273.15+A8)	=m*C8+b		**18.02**
9	=A8	=(1-1/(1+1000/mw))*B8	=C8	=LN(B9)	**<- 1 molal**	
10			=D9-m*C9	new	constant	
11			=m/(LN(760)-C10)	b.p.	T(K)	
12			=C11-273.15	b.p.	t(C)	
13			**=C11-H5**	**<- Kb**	C/molal	

Worksheet 13.5F Formulas for Worksheet 13.5

Formula List 13.5 Definitions and Formulas for Worksheet 13.5

Cell E5	Defined as m
Cell F5	Defined as b
Cell G5	=-0.0083145*m
Cell H5	=m/(LN(760)-b)
Cell I5	=H5-273.15
Cell F8	Defined as mw

Problems

1. What is the enthalpy of vaporization of benzene at its boiling point? What is the boiling point of benzene at atmospheric pressure? What is the vapor pressure of benzene at 20°C?

2. What is the boiling point of toluene at atmospheric pressure? What is the enthalpy of vaporization of toluene at its boiling point? What is the vapor pressure of toluene at 23°C?

3. What is the boiling point of carbon disulfide at atmospheric pressure? What is the enthalpy of vaporization of carbon disulfide at its boiling point? What is the vapor pressure of carbon disulfide at 10°C?

4. What is the boiling point of boron tribromide at atmospheric pressure? What is the enthalpy of vaporization of boron tribromide at its boiling point? What is the vapor pressure of boron tribromide at 15°C?

5. What is the boiling point of acetone at atmospheric pressure? What is the enthalpy of vaporization of acetone at its boiling point? Calculate the vapor pressure of acetone at 23°C.

6. What is the boiling point of chloroform and methanol at atmospheric pressure? What is the enthalpy of vaporization of chloroform and methanol at its boiling point? What is the vapor pressure of chloroform and methanol at 10°C? Which has the higher boiling point? Which has the higher vapor pressure at 10°C?

7. On the same graph, plot the vapor pressure of acetone, benzene, and water versus the temperature.

8. On the same graph, plot the natural log of the vapor pressure versus the inverse of the temperature for ethanol, water, and toluene.

9. Plot the partial pressures and the total pressure of a solution of chloroform and toluene as a function of the mole fraction of chloroform. Use a temperature of 30°C. Do another plot at a temperature of 60°C.

10. What is the vapor composition above a solution of chloroform and toluene when the liquid phase has 0.6 of a mole fraction of chloroform? Calculate the vapor composition at both 30°C and 60°C.

11. Plot both a "liquid" and a "vapor" line for a solution of chloroform and toluene at 60°C. Use mole fraction of chloroform on the abscissa or horizontal axis.

12. What is the molal boiling-point-elevation constant for carbon disulfide? What is the boiling point of the pure liquid at atmospheric pressure?

13. What is the boiling point of pure boron tribromide at atmospheric pressure? What is the molal boiling-point-elevation constant for boron tribromide?

14. What is the boiling point of a 4.0 molal solution of biphenyl in benzene? Consider biphenyl to be nonvolatile at this temperature.

15. What is the molal boiling-point-elevation constant for acetone?

14

Chemical Kinetics

Chemical kinetics is concerned with the rates at which chemical reactions proceed. Researchers who study reaction rates are not only interested in the speed with which reactions occur, but with the variables that affect these rates. A variable of primary concern is the concentration of various species and the effect that these concentrations have on the reaction rate. This relationship must be determined experimentally; the stoichiometry of the reaction may or may not be indicative of the mathematical rate expression. The rate of reaction is expressed as the change in molar concentration of a reactant or product per unit time. A rate law represents this change in molar concentration per unit time as a function of a rate constant, k, and the molar concentrations of the pertinent species that affect the rate. The molar concentrations as expressed in the rate law are raised to an appropriate power by an exponent that is frequently an integer, although non-integer values are possible. If a rate law has only a single concentration term and the exponent of that concentration term is unity, then the rate law is designated as first-order. When the rate law has a single concentration term and the exponent of that term is two, then the rate law is designated as second-order. The mathematical and graphical description of first-order reactions will be our first consideration. From this we will expand upon the difference between an average rate of change and an instantaneous rate of change. Next, the mathematical consequences of second-order kinetic behavior will be illustrated with an emphasis on a comparison of the rate with which second-order reactions approach completion as contrasted to that of first-order reactions. Spreadsheets provide a very convenient format for assessment of experimental kinetic data. This capability is utilized for the analysis of kinetic data for the reaction between ammonium ion and nitrite ion. A second example of the evaluation of initial rate data is illustrated by the reaction of peroxydisulfate ion with iodide ion. An important consideration of any kinetics experiment is the temperature dependence of the reaction rates and the corresponding rate constants. The computation of the rate constant at any specified temperature is readily accomplished with spreadsheets with emphasis on competitive reactions. (BLB Chap. 14)

First-Order Kinetics

The behavior of many chemical reactions can be described by first-order kinetics. In addition, two physical phenomena, radioactive nuclear decay and fluorescence, are also governed by first-order kinetics. All nuclear decay reactions are first-order with very large variations in the first-order half-lives having very important physical and chemical consequences. The decay in the fluorescent intensity after the excitation of a species that exhibits fluorescence is governed by first-order kinetics. A first-order reaction has the following functional form.

$$[A]_t = [A]_{t=0}e^{-kt}$$

The preceding equation is the integral form (normal algebraic form) of the differential equation that represents the rate of change of reactant per unit time as a function of the rate constant and the concentration of the reactant. The differential form of the equation describing first-order kinetics is

$$-d[A]_t/dt = k[A]_t$$

This equation describes the continually decreasing rate of disappearance of the reactant as the concentration of that reactant decreases. This differential form of the equation is the slope (tangent) of the algebraic equation at any chosen time. This slope has a large negative value during the initial stages of the reaction and subsequently decreases until the value is zero.

The half-life of a first-order reaction is the time necessary for half of the remaining material to react. This particular mathematical function that describes first-order kinetics has the property of requiring the same interval of time to reduce $[A]_0$ to one-half of its initial value, regardless of the initial numerical value. The half-life is independent of the initial concentration and it is calculated by

$$t_{1/2} = (\ln 2)/k$$

The extent of reaction is frequently expressed in terms of the percentage of reactant remaining after a stated period of time or as the number of half-lives that have elapsed. Worksheet 14.1 illustrates an example of first-order kinetics with a carefully selected rate constant. The value of k, the first-order rate constant, may be varied as well as the initial concentration of the reactant, A. On this spreadsheet, the time intervals are calculated as the appropriate half-lives as determined by k. These may be altered to include a longer or shorter period of time, although you will observe that more than 99% of the reaction has taken place in seven half-lives. The formulas for producing the calculations presented by this worksheet are given in Formula List 14.1, which follows the presentation of this spreadsheet.

	A	B	C	D	E
1	First-Order Kinetics		k-->	0.069314	1/s
2			A(initial)-->	1	M
3	time (s)				
4		[A]	% A remaining		
5	0.0	1	100.00%		
6	10.0	0.5	50.00%		
7	20.0	0.25	25.00%		
8	30.0	0.125	12.50%		
9	40.0	0.0625	6.25%		
10	50.0	0.03125	3.13%		
11	60.0	0.015625	1.56%		
12	70.0	0.0078125	0.78%		
13					
14	time (s)	[A]	% A remaining	t(1/2) elapsed	
15	15	0.354	35.36%	1.50	

Worksheet 14.1 First-Order Kinetics

The formulas that compute [A] and the percent of A remaining after an indicated time occupy column B and column C. The contents of cell A7 are copied into cells A8 through A12 by using the Fill Down command under Edit on the menu bar. In a similar fashion, the contents of cell B5 are copied into cells B6 through B12 using the Fill Down command. The contents of cell C5 are copied into cells C6 through C12.

Formula List 14.1 Formulas for Worksheet 14.1

Cell A6	=LN(2)/D1	$(\ln 2)/k$
Cell A7	=A6+$A$6	$t_i + t_{1/2}$
Cell B5	=$D$2*EXP(-$D$1*A5)	$[A]_{t=0}e^{-kt}$
Cell C5	=B5/$D$2	$[A]_t/[A]_{t=0}$
Cell D15	=A15/A6	$t/t_{1/2}$

A plot of the relationship between the concentration of reactant, *A*, and time, *t*, is illustrated by Chart 14.1. This graphical presentation of first-order kinetic behavior does not appear as a smooth function due to the limited number of data points. A detailed graph (a graph produced by plotting many more points which reduces the time interval between the calculated points) of first-order kinetics produces a smooth curve with a continuing decrease in the rate of disappearance of the reactant, *A*. The initial rate of disappearance of the reactant is at a maximum value at $t = 0$, and thereafter the rate of disappearance of the reactant decreases in a smooth, continuous manner until the slope is zero. At the point at which the slope is zero, $-d[A]_t/dt = 0$, the reaction is complete or at equilibrium, which indicates that the forward and reverse rates are equal.

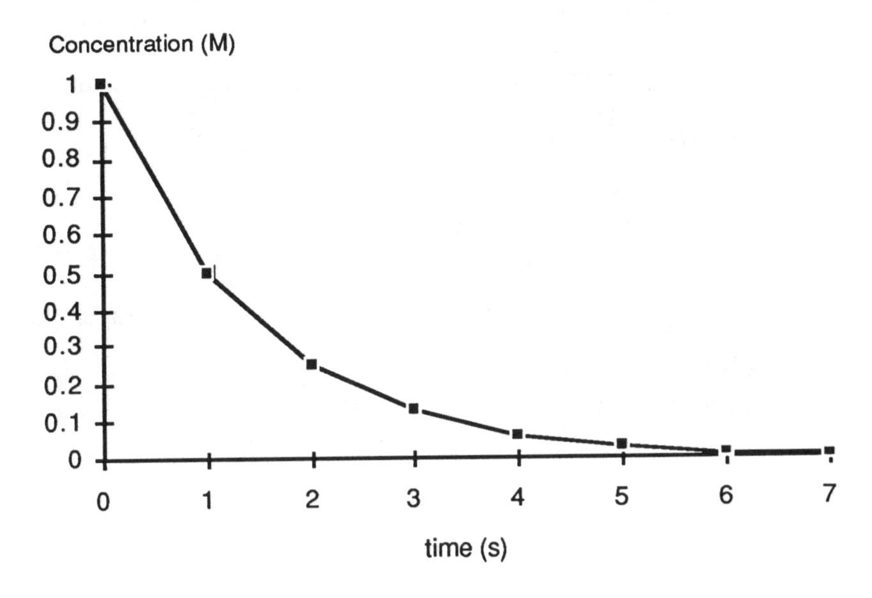

Chart 14.1 First-Order Kinetics

Instantaneous Rate of Change

The previous plot of a first-order reaction delineates the problem inherent in portraying a nonlinear function with a limited number of data points. When a series of wide intervals are represented by pairs of points connected by straight lines, the resulting graph of the nonlinear function suffers discontinuities at the junction of every pair of lines. The results are most dramatic during the early segments of the graph due to the large changes in the rate of reaction. The nonlinear function is in reality a smooth, continuous curve that may be observed by plotting a substantial number of points. The large number of plotted points has the effect of reducing the size of the intervals that are connected by straight lines.

Another problem in the treatment of kinetic data is the need for calculating the instantaneous rate of disappearance of a reactant or the instantaneous rate of appearance of a product. The numerical value of the instantaneous rate of change of the species is represented by the slope of the tangent to the curve at a given time, t. The first derivative of a function, evaluated at a given value of *t*, represents the instantaneous slope or the instantaneous rate of change of the concentration at the specified time. The instantaneous rate of change of the concentration at a specified time is represented by $-d[A]_t/dt$, and the average rate of change of the concentration over an interval is represented by $-\Delta[A]_t/\Delta t$. The crucial aspect is the difference between the instantaneous slope at a selected point on the curve and the slope as determined by a finite interval. This finite interval is bounded by one point on the curve and another point that is produced by a linear projection. With a nonlinear function, the second point cannot lie on the curve. The average rate of change of a function approaches the instantaneous rate of change of the function as the interval, Δt -> 0.

$$\lim_{\Delta t \to 0} -\Delta[A]_t/\Delta t = -d[A]_t/dt$$

The difference between the instantaneous rate of change and the average rate of change is a function of the size of the chosen interval for the calculation of the average rate of change. This difference over an interval of one half-life for an initial concentration of 1.0 M is illustrated in Chart 14.2. The difference can be reduced substantially by reducing the size of the interval.

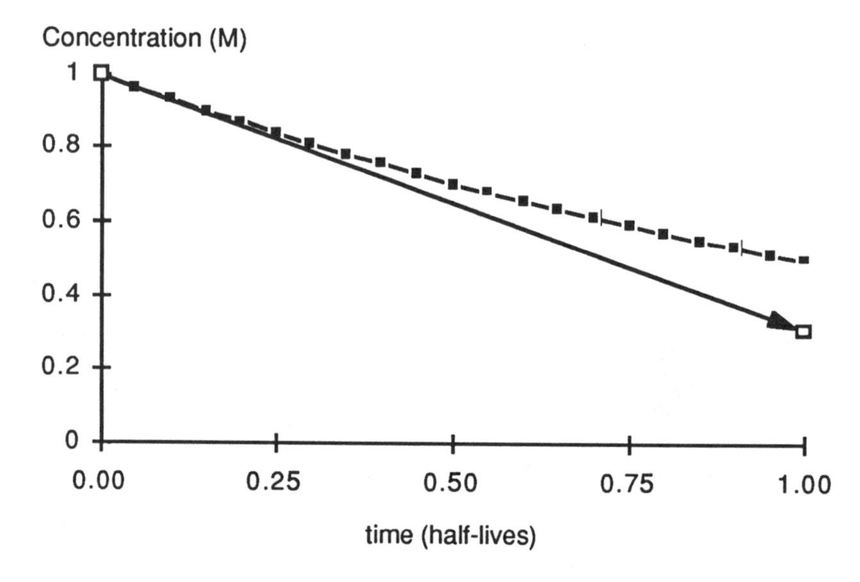

Chart 14.2 Instantaneous and Average Rate of Change

Calculations of the real and approximate [*A*], and the average and instantaneous rate of change of the reactant are shown as Worksheet 14.2.

This spreadsheet may be used to investigate the consequences produced by changing the size of the time interval. The spreadsheet is designed to calculate ten equal-size time intervals where the size of the time interval is determined by multiplying the chosen decimal interval by the half-life of the reaction. If a value of 0.1 is chosen for the interval, the total length of time considered is one half-life. If an interval of 1.0 is used, then the total time will be ten half-lives. The equations used for calculating the variables in a particular column are presented in Formula List 14.2.

Formula List 14.2 Average and Instantaneous Rate of Change

Column B	$[A]_t = [A]_{t=0}e^{-kt}$
Column C	$[A]_t = [A]_{t-1} + (\Delta[A]_{t-1}/\Delta t)\cdot\Delta t$
Column D	$-\Delta[A]_t/\Delta t = k[A]_t$
Column E	$-d[A]_t/dt = k[A]_t$

The instantaneous rate of change of the reactant at a selected time is calculated in Column E. The values of the instantaneous rate of change are negative, as noted by the parentheses. As the reaction proceeds with an increase in t, each subsequent value of the instantaneous rate of disappearance of the reactant decreases. Thus, each value of t results in a unique and different value of $-d[A]_t/dt$. The calculation for the approximate rate of disappearance of the reactant assumes that the rate of change of the concentration per unit time (the slope) is constant over the specified time interval. This is not correct except in the limit as $\Delta t \to 0$.

	A	B	C	D	E
5		[A] real	[A] approx	Av Rate	Ins Rate
6	0	=$D$2*EXP(-$D$1*A6)	=D2	=$D$1*C6	=$D$1*B6
7	=$D$3*LN(2)/D1	=$D$2*EXP(-$D$1*A7)	=C6-D6*$D$3	=$D$1*C7	=$D$1*B7

Worksheet 14.2F Average and Instantaneous Rate of Change

Second-Order Kinetics

A reaction with a single reactant that proceeds by second-order kinetics is governed by the following differential equation:

$$-d[A]_t/dt = k[A]_t^2$$

The integral or algebraic form of the equation that describes the relationship between the reactant concentration and the time variable for a reaction that is second-order in the single reactant is

$$\frac{1}{[A]_t} = \frac{1}{[A]_{t=0}} + kt$$

A plot or curve described by this function is very similar to the curve that is observed for first-order reactions except for the subtle decrease in the rate of reaction as the time variable increases. In general, a second-order reaction is a much slower reaction than a reaction governed by first-order kinetics when both reactions have identical initial concentrations. This comparison is valid for first-order and second-order reactions that have the same numerical value for their respective rate constants and that have an initial reactant concentration of one molar or less.

A half-life expression for second-order kinetics is dependent upon the chosen initial concentration. This variable half-life is the time interval required for a twofold decrease in the chosen initial concentration of the reactant. The half-life relationship for a second-order reaction is

$$t_{1/2} = \frac{1}{k[A]_{t=0}}$$

At subsequent stages in the reaction, the half-life increases due to the decreasing concentration of the reactant, $[A]$. The choice of an initial concentration or time variable is arbitrary. This choice, however, then establishes a half-life time interval associated with this initial value. It is important to realize that the half-life is not constant for a second-order reaction as contrasted with the invariable half-life for a first-order reaction. A graph of second-order kinetics is illustrated as Chart 14.3.

A spreadsheet that calculates for second-order kinetics the concentration of the reactant as a function of time and the percent of reactant remaining is illustrated as Worksheet 14.3. This spreadsheet computes a constant time interval that is based on a second-order half-life as determined by the reciprocal of the product of the second-order rate constant, k, and the initial concentration of the reactant, $[A]_{t=0}$. The spreadsheet also computes the concentration of the reactant and the percent of reactant remaining for any time chosen by the user. The crucial formulas for producing the computations executed by this worksheet are given in Formula List 14.3, which follows the presentation of this spreadsheet.

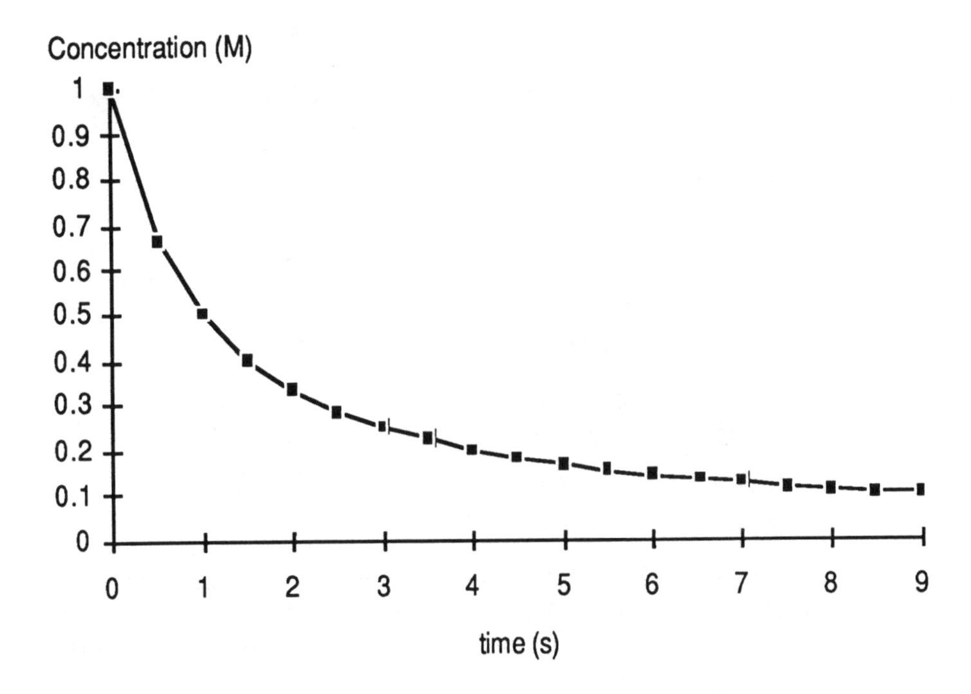

Chart 14.3 Second-Order Kinetics

	A	B	C	D	E
1	Second-Order Kinetics		k-->	1	1/M-s
2			A(initial)-->	1	M
3	time (s)				
4		[A]	% A remaining		
5	0.0	1.000	100.0%		
6	1.0	0.500	50.0%		
7	2.0	0.333	33.3%		
8	3.0	0.250	25.0%		
9	4.0	0.200	20.0%		
10	5.0	0.167	16.7%		
11	6.0	0.143	14.3%		
12	7.0	0.125	12.5%		
13	8.0	0.111	11.1%		
14	9.0	0.100	10.0%		
15	10.0	0.091	9.1%		
16					
17	time (s)	[A]	% A remaining		
18	20	0.048	4.76%		

Worksheet 14.3 Second-Order Kinetics

The formulas used to compute [A] and the percent of A remaining after a specified time are presented in Formula List 14.3. The contents of cell A7 are copied into cells A8 through A15 by using the Fill Down command under Edit on the menu bar. In a similar fashion, the contents of cell B5 are copied into cells B6 through B15 using the Fill Down command. The contents of cell C5 are copied into cells C6 through C15.

Formula List 14.3 Formulas for Worksheet 14.3

Cell A6	=1/(D1*D2)	$1/k[A]_{t=0}$
Cell A7	=A6+$A$6	$t_i + t_{1/2}$
Cell B5	=(1/$D$2+$D$1*A5)^-1	$(1/[A]_{t=0}+kt)^{-1}$
Cell C5	=B5/$D$2	$[A]_t/[A]_{t=0}$

Concentration data for second-order kinetics can be graphed in a manner that produces a linear plot. The inverse of the concentration of the second-order term is used as the vertical axis and time is plotted on the horizontal axis. The slope of this curve is the second-order rate constant in units of reciprocal concentration and reciprocal time. This type of graph is illustrated as Chart 14.3B, which was produced from the previous data.

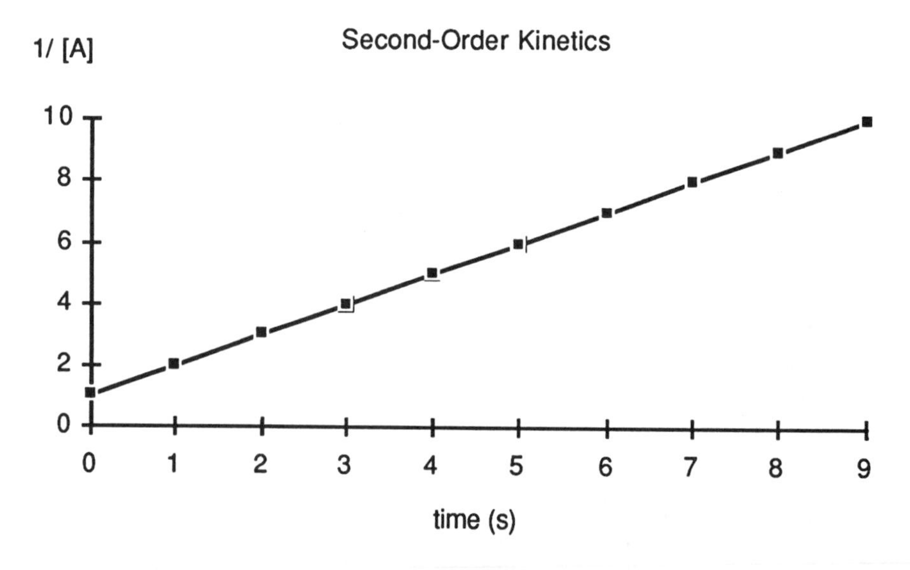

Chart 14.3A Linear Plot of Second-Order Kinetics

Comparing First-Order and Second-Order Kinetics

A reaction that is governed by second-order kinetics is much slower in achieving completion or equilibrium than a reaction that is governed by first-order kinetics when both have the same numerical value for the rate constant. A second-order reaction requires an increasing interval of time to complete a specified fraction of reaction (e.g., reduction to one-half the present concentration) as the reaction progresses toward completion. Worksheet 14.4 presents a comparison of the rate with which second-order reactions approach completion as contrasted to that of first-order reactions. A constant time interval that is based on the half-life of the first-order reaction is used. The percent of reactant remaining for a specified time for both types of kinetic behavior is then computed for a total time period that represents ten half-lives for the first-order reaction. The second-order reaction is based on a single component reaction.

	A	B	C	D	E
1	First/Second-Order Kinetics		k-->	0.69314	
2			A(initial)-->	1	M
3					
4	time (s)	% A remaining			
5		First-order	Second-order		
6	0.0	100.00%	100.00%		
7	1.0	50.00%	59.06%		
8	2.0	25.00%	41.91%		
9	3.0	12.50%	32.47%		
10	4.0	6.25%	26.51%		
11	5.0	3.13%	22.39%		
12	6.0	1.56%	19.38%		
13	7.0	0.78%	17.09%		
14	8.0	0.39%	15.28%		
15	9.0	0.20%	13.82%		
16	10.0	0.10%	12.61%		

Worksheet 14.4 Comparing First/Second-Order Kinetics

The formulas used in this spreadsheet that compares the percent of A remaining after a specified time for both first-order and second-order kinetics are presented in Table 14.4. The contents of cell A8 are copied into cells A9 through A16 by using the Fill Down command under Edit on the menu bar. In a similar fashion, the contents of cell B6 are copied into cells B7 through B16 using the Fill Down command. The contents of cell C6 are copied into cells C7 through C16.

Formula List 14.4 Formulas for Worksheet 14.4

Cell A7	=LN(2)/D1	$(\ln 2)/k$
Cell A8	=A7+$A$7	$t_i + t_{1/2}$
Cell B6	=EXP(-$D$1*A6)	$[A]_{t=0}e^{-kt}/[A]_{t=0}$
Cell C6	=(1/$D$2+$D$1*A6)^-1/$D$2	$((1/[A]_{t=0}+kt)^{-1})/[A]_{t=0}$

A graph comparing the progress of a first-order and second-order reaction is illustrated as Chart 14.4. These results dramatically illustrate that the first-order reaction approaches completion in a much more rapid fashion than a second-order reaction. When the curves are viewed simultaneously, these consequences are apparent, but when viewed independently it is not obvious whether a curve represents first or second-order kinetics. The increase in the half-life of second-order kinetics at subsequent stages of the reaction due to the decreasing concentration of the reactant is apparent from this graph.

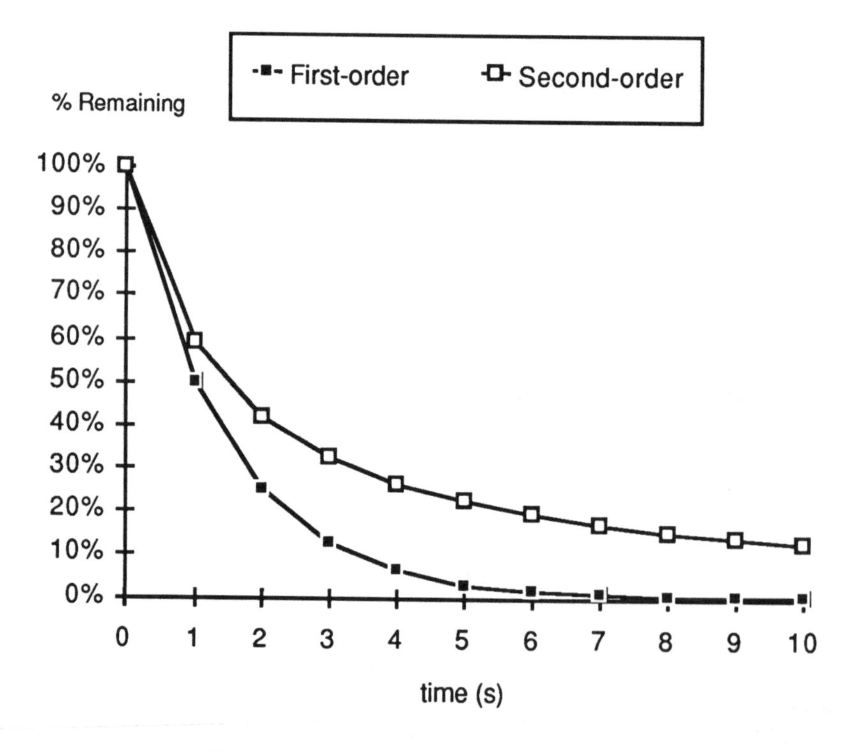

Chart 14.4 First/Second-Order Kinetics

Analysis of Kinetic Data

Spreadsheets provide a very convenient format for the evaluation of experimental kinetic data (Chapter 23). Second-order reactions of the following type are among the most common of all reactions studied.

$$-\mathrm{d}[A]_t/\mathrm{d}t = k[A]_t \cdot [B]_t$$

A general solution of this differential equation based on the stoichiometry of the chemical reaction produces a rather complicated equation. Computation of the results is greatly simplified if the experimental conditions allow one of the reactant species to be initially present at approximately ten times the concentration of the other species. Under this set of conditions, the reaction can be mathematically treated as if it is a first-order reaction. If the experimental conditions allow, $[B]_{t=0} = 10 \cdot [A]_{t=0}$, then the second-order differential equation becomes

$$-\mathrm{d}[A]_t/\mathrm{d}t = k[B]_{t=0} \cdot [A]_t$$

This situation is known as pseudo first-order kinetics, and the algebraic form of the equation is

$$[A]_t = [A]_{t=0} e^{-k[B]_{t=0} t}$$

The concentration of species, B, will vary by ten percent during the entire reaction when the initial concentration of B is ten times that of A. The conservation of mass equation, derived from the chemical equation, can be used to correct for the slight change in the concentration of B. In this case, the chemical reaction has a one-to-one stoichiometry.

$$[B]_t = [B]_{t=0} - ([A]_{t=0} - [A]_t)$$

This equation is a general equation that is valid for any pair of selected times, as the difference between the concentration of a species for the two times represents the amount of that species that reacted during the chosen time interval. This experimental approach provides a method for evaluating the second-order rate constant and concentrations of the species as the reaction proceeds for this important class of reactions.

The Ammonium and Nitrite Ion Reaction

The electron transfer reaction between ammonium ion and nitrite ion in aqueous solutions is first-order in each reactant and thus second-order overall. Worksheet 14.5 illustrates the analysis of initial rate data for the second-order reaction between ammonium ion and nitrite ion. Initial rate data for this reaction can be found in *CHEMISTRY The Central Science*, 5th Edition, by Brown, LeMay, and Bursten.

	A	B	C	D	E
1	Second-Order Kinetics		NH4+ + NO2- -> N2 + 2H2O		
2					
3	time(s)				
4		[NH4+]	[NO2-]	Instan Rate	k(1/M-s)
5	0	0.0200	0.200	1.07E-06	2.68E-04
6	9000	0.0124	0.192	6.36E-07	
7	18000	0.0079	0.188	3.98E-07	
8	27000	0.0051	0.185	2.55E-07	
9	36000	0.0034	0.183	1.65E-07	
10	45000	0.0022	0.182	1.07E-07	
11	54000	0.0014	0.181	6.98E-08	
12	63000	0.0009	0.181	4.55E-08	
13	72000	0.0006	0.181	2.96E-08	
14	81000	0.0004	0.180	1.93E-08	
15	90000	0.0003	0.180	1.25E-08	

Worksheet 14.5 Reaction of Ammonium Ion and Nitrite Ion

Formula List 14.5 presents the important formulas that are used and copied into the cells in Worksheet 14.5. The approach used in this spreadsheet is only valid when one of the reactants has a high initial concentration relative to the other reactant. As the reaction proceeds, the concentration of the reactant that is initially present at the lower concentration can be approximated by the solution for first-order kinetics. It is assumed that the concentration of the second species remains constant, but in reality the expression for the conservation of mass allows computation of these minor changes in concentration.

Formula List 14.5 Formulas for Worksheet 14.5

Cell A6	=A5+9000
Cell B6	=$B$5*EXP(-$E$5*C5*A6)
Cell C6	=C5-(B5-B6)
Cell D6	=$E$5*B6*C6
Cell E5	=D5/(B5*C5)

The contents of cell A6 are then copied into cells A7 through A15. In a similar fashion, this pattern is then repeated for the contents of cells B6, C6, and D6. The initial concentrations of the two reactant species and the initial rate can then be altered in row five and the corresponding computations made with the restriction that the initial concentration of one of the species has to be high relative to the other so that a pseudo first-order treatment of the data is valid. A graph of the concentrations of the two species as a function of time under the conditions presented in Worksheet 14.5 is presented as Chart 14.5.

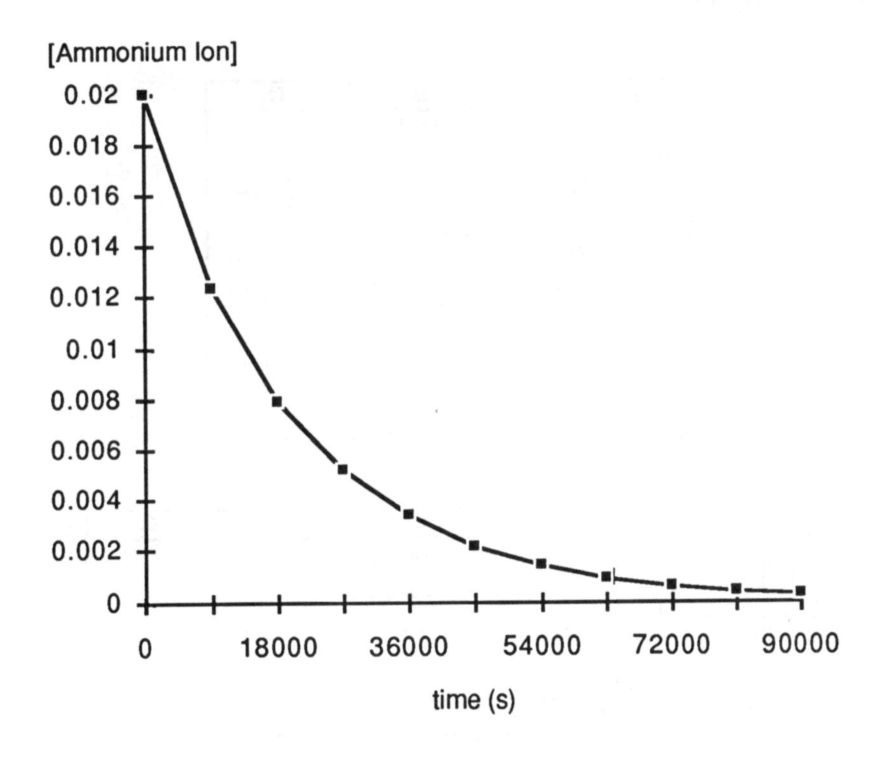

Chart 14.5 Reaction of Ammonium and Nitrite Ion

Temperature Dependence of the Rate Constant

The rate of a chemical reaction increases as the temperature increases. One of the ways that chemists express this temperature relationship is by the Arrhenius equation

$$\ln(k_1/k_2) = E_a(1/T_2-1/T_1)/R$$

where R is the ideal-gas constant of 8.314 joules. This temperature relationship has some interesting consequences.

A rule of thumb that is valid for several chemical reactions is that the rate of reaction doubles when the temperature is increased by ten degrees Celsius. Strictly speaking, this is only true for reactions that have an activation energy of 51.506 kJ when the temperature is raised ten degrees from an initial temperature of 294 K. Worksheet 14.6 illustrates the validity of this statement. This spreadsheet can be used to calculate the rate constant at any temperature when the activation energy and rate constant are known at a given temperature.

	A	B	C
1	Temperature Dependence		
2			
3	k (T1)	1	<- Input
4	T1 (K)	294	<- Input
5	T2 (K)	304	<- Input
6	Ea (kJ)	51.506	<- Input
7			
8	k (T2)	2.00E+00	

Worksheet 14.6 Temperature Dependence for Rate Constants

	A	B	C
1		Temperature Dependence	
2			
3	k (T1)	1	<- Input
4	T1 (K)	294	<- Input
5	T2 (K)	304	<- Input
6	Ea (kJ)	51.506	<- Input
7			
8	k (T2)	=B3/EXP(1000*B6*(1/B5-1/B4)/8.314)	

Worksheet 14.6F Formulas for Worksheet 14.6

With the rate of a chemical reaction increasing as the temperature increases, the activation energy is always positive for all chemical reactions. This single direction behavior leads to an important consequence when competitive reactions are considered. When considering more than one reaction, the reaction with the largest activation energy is most sensitive to changes in temperature. This result can be observed by calculating the new rate constants for two different reactions that have very different numerical values for their activation energies. The reaction with the largest activation energy undergoes the most dramatic increase in its rate constant when the temperature is increased. With two competing reactions, this aspect can be exploited to control the relative yields of product of the two reactions. Worksheet 14.7 illustrates that a 50-degree change in temperature results in a dramatic reversal in the reaction that predominates in this competitive situation. For this calculation, reaction A has $k = 0.1$ and $E_a = 52$ kJ, and reaction B has $k = 0.6$ and $E_a = 12$ kJ. At the lower temperature, reaction B is six times as fast as reaction A. But at the higher temperature, the reverse is observed with reaction A being almost twice as fast as reaction B.

	A	B	C
1	**Temperature Dependence**		
2			
3	T1 (K)	294	
4	T2 (K)	344	
5			
6		reaction A	reaction B
7	Ea (kJ)	52	12
8	k (T1)	**0.1**	**0.6**
9	k (T2)	**2.20**	**1.22**

Worksheet 14.7 Competitive Reactions

Problems

1. The linkage isomer methyl isonitrile converts to acetonitrile with a first-order rate constant of 5.11×10^{-5} s^{-1}. After 20,000 s if 36% of the initial 0.100 M reactant remains, is this consistent with a first or second-order reaction? After what time interval will 12.5% remain? If the reaction were second-order, what would the rate constant be?

2. The copper(II) catalyzed reaction of vanadium(III) and several cobalt(III) complexes is pseudo first-order. The first-order rate constant is 0.0347 s^{-1} when the copper(II) concentration is 0.0982 M. What is the half-life? When will 3.12% remain? When will less than 1% remain, and at the end of that time interval, how much reactant would be present if the reaction were second-order?

3. With the rearrangement of methyl isonitrile to acetonitrile, $k = 2.52 \times 10^{-5}$ s^{-1} at 189.7 ° C and $E_a = 153.6$ kJ, what is the rate constant at 230.3°C?

4. An organic reaction has two different paths, A and B, by which two different products are produced. At 400.9 K the rate constants are identical, $k = 1.04 \times 10^{-2}$. If the activation energy of path A is 118 kJ and path B is 84.3 kJ, what are the respective rate constants at a temperature of 341.3 K? Which is faster and by what factor? At 485.8 K which reaction is the fastest and by what factor? From this example, make some general conclusions about the temperature dependence of reaction rates.

5. The decomposition of N_2O_5 has a rate constant of 7.87×10^{-7} s^{-1} at a temperature of 273 K. With an activation energy of 103 kJ, what is the rate constant at 298 K? What is the rate constant at 308 K?

15

Chemical Equilibrium

Most chemical reactions proceed to a point where they appear to stop. Analysis at this point shows that the reaction mixture contains both products and reactants. No further changes in the concentrations are observed, regardless of the time allowed. Systems that have achieved this apparent static state are said to be at chemical equilibrium. The conditions under which the reaction is carried out determine what the reactant and product concentrations are when the system has reached equilibrium. The position of equilibrium can be altered after the reaction has appeared to cease by varying one or more of the parameters that affect the equilibrium position. It has been concluded from this and other evidence that a system at equilibrium continues to react in both the forward and reverse directions, but at equal rates. The reaction conditions of most interest to the chemist are those of temperature, total pressure, and the concentrations of all species. In this chapter, we examine how altering reaction conditions affects these concentrations, and we solve equations for the position of equilibrium. Changes in the equilibrium position brought about by changes in the reaction conditions such as temperature, concentration, and pressure are collectively described by Le Chatelier's Principle. (BLB Chap. 15)

The Equilibrium Constant

From both an experimental and theoretical basis, it has been observed that a system at equilibrium may be described in a mathematical form that produces a constant regardless of the equilibrium concentrations. This constant is a function of the temperature, and it is slightly altered by the nature of the other species that are present, although we will ignore this aspect of the equilibrium constant. A generalized chemical reaction may be represented by italicized letters for the stoichiometric coefficients and capital letters for the chemical species:

$$h\text{A} + i\text{B} = j\text{P} + k\text{S}$$

For this chemical reaction, a mathematical relationship may be written that expresses the ratio of the product concentrations to the reactant concentrations, each raised to an appropriate power. This relationship has the following form where the convention is to place the products in the numerator and the reactants in the denominator:

$$K_{eq} = \frac{[P]^j[S]^k}{[A]^h[B]^i}$$

The square brackets signify the molar concentrations of the species. The equilibrium constant, K_{eq}, is essentially a constant for a given reaction at a specified temperature. A similar relationship may be written in terms of the partial pressures of the species when gases are involved. It is customary to use a subscript "p" on the equilibrium constant to signify that partial pressures are being used instead of molar concentrations. The numerical value of the equilibrium constant depends upon the choice of the concentration units. It should be emphasized that the partial pressures or concentrations used in this relationship are those that are measured when the system is at equilibrium.

For a reaction such as the equilibrium between the dimer and monomer, dinitrogen tetraoxide and nitrogen dioxide,

$$N_2O_4\ (g) = 2\ NO_2\ (g)$$

the experimentally measured equilibrium constant at 100°C is 0.212. This reaction has the following equilibrium expression:

$$K_{eq} = \frac{[NO_2]^2}{[N_2O_4]}$$

For this mathematical relationship, there exists an infinite number of equilibrium concentrations that will satisfy this equilibrium expression. It is necessary that the equilibrium concentrations be measured when the system is at equilibrium and that these measurements do not disturb this equilibrium state. Substitution of appropriate experimental equilibrium concentrations into the equilibrium expression will yield a constant within experimental error for the temperature at which the measurements where made.

Nonequilibrium concentrations, when substituted into the equilibrium expression, produce a value that does not equal K. There are an infinite number of concentrations that do not satisfy the equilibrium expression, and these represent reaction concentrations at nonequilibrium conditions. If a system is not at equilibrium, it will proceed in the direction necessary to achieve equilibrium. For nonequilibrium conditions, it is customary to use Q, called the reaction quotient. The value of Q is calculated in a manner that is completely analogous with the calculation of the equilibrium constant, except the concentrations or partial pressures may or may not be equilibrium values. The value of Q may be viewed as a trial value to ascertain whether the system is at equilibrium. If the system is not at equilibrium, the value of Q signifies the direction in which the reaction will proceed in order to achieve equilibrium. If $Q > K$, the ratio of products to reactants is too large and the reaction position will shift toward the reactants. The decrease in the concentration of the products and the corresponding increase in the concentration of the reactants continues until the concentrations are such that $Q = K$ and the system is at equilibrium. If $Q < K$, the system contains concentrations of reactants that are too high and concentrations of products that are too low. In this case, the reaction position shifts toward the products.

Equilibrium Concentrations

The equilibrium constant can be used to determine the resulting equilibrium concentrations when the initial concentrations are known. The dinitrogen tetraoxide and nitrogen dioxide equilibrium will be used as an example of this calculation.

$$N_2O_4\ (g) = 2\ NO_2\ (g)$$

	N_2O_4	NO_2
Initial	D	M
Change	$-x$	$2x$
Equilibrium	$D - x$	$M + 2x$

All of the values given represent concentrations of mol/L or partial pressures in atm. If the reaction position happens to adjust in the opposite direction, then the value of x is negative and this representation is still correct.

$$K_{eq} = \frac{[NO_2]^2}{[N_2O_4]} = \frac{(M + 2x)^2}{D - x}$$

$$4x^2 + (4 \cdot M + K)x + M^2 - K \cdot D = 0$$

$$x = \frac{-(4 \cdot M + K) + \sqrt{(4 \cdot M + K)^2 - 16 \cdot (M^2 - K \cdot D)}}{8}$$

Only concentrations that are positive have a physical meaning. There are two solutions, but only the one with the positive discriminant produces positive concentrations of dinitrogen tetraoxide and mononitrogen dioxide. This is true because x must be positive when $M = 0$ in order to have a positive concentration of $[NO_2]$. In a similar fashion, x must be negative when D = 0 in order to have a positive concentration of $[N_2O_4]$. These computations are produced by Worksheet 15.1, which is not shown.

	A	B	C	D
1	N2O4 =	2 NO2	Equilibrium	
2	**Keq =**	**0.212**	="100 C"	
3				
4		Initial	Equilibrium	
5	[N2O4] =	0.212	=D-(-(4*M+K)+((4*M+K)^2-16*(M^2-K*D))^0.5)/8	
6	[NO2] =	0.212	=M+(-(4*M+K)+((4*M+K)^2-16*(M^2-K*D))^0.5)/4	
7				
8	**Q =**	**=B6^2/B5**	**=C6^2/C5**	**= Keq**
9			=IF(Q=K,"equilibrium","")	
10			=IF(Q>K,"Q>K shift toward reactants","")	
11			=IF(Q<K,"Q<K shift toward products","")	

Worksheet 15.1F Equilibrium Concentrations

Worksheet 15.1F illustrates the formulas used for calculating the direction in which the reaction will proceed and the resulting equilibrium concentrations. The initial concentrations of dinitrogen tetraoxide and nitrogen dioxide can be entered in cells B5 and B6. The spreadsheet will calculate Q and note the direction of reaction when the system is not at equilibrium. The final equilibrium concentrations are computed, and cell C8 recalculates the value of the equilibrium constant from these values. Checking the equilibrium concentration values by recomputing the equilibrium constant serves as an assurance that the spreadsheet functions as intended. LeChatelier's Principle provides predictions on the effect of changing the concentrations of reactants and products, and the computed equilibrium concentrations verify the correctness of this principle.

The Effect of Pressure

The previous spreadsheet, which is titled Worksheet 15.1, can be extended so that the partial pressures, the total pressure, and the equilibrium constant, K_p, can be computed. For our purposes, we can consider that the concentration equilibrium constant and the partial pressure equilibrium constant represent the same experimental value expressed in different units. The equation for converting one to the other is

$$K_p = K_c(R \cdot T)^{\Delta n}$$

The following changes will produce Worksheet 15.2F, which will compute the initial partial pressures and equilibrium partial pressures for any chosen initial molar concentrations.

	A	B	C	D
13	Pressure	atm		
14	P (N2O4) =	=0.08205*373.2*B5	=0.08205*373.2*C5	
15	P (NO2) =	=0.08205*373.2*B6	=0.08205*373.2*C6	
16	P (total) =	=B14+B15	=C14+C15	
17				
18			=C15^2/C14	= Kp

Worksheet 15.2F Concentrations and Partial Pressures

The results computed in cell C18 should be identical with the value of Kp as calculated by the equation that converts K_c. You should notice that in those cases where the initial concentrations produce a value of $Q < K$, the reaction proceeds to increase the concentrations of the products and to concurrently reduce the concentrations of the reactants until $Q = K$. These cases produce final equilibrium concentrations that result in a higher total pressure than the initial concentrations. This phenomenon will be considered in more detail in Chapter 20, Free Energy and Equilibrium.

A more interesting approach for understanding the effects of pressure on the ultimate equilibrium position is provided by designing a spreadsheet that calculates the equilibrium partial pressure for the component gases when a total pressure is applied to the system. The system will adjust the total internal pressure so that it is equal to the applied external pressure.

$$K_p = \frac{P^2_{NO_2}}{P_{N_2O_4}} = \frac{P^2_M}{P_D}$$

$$P_T = P_{N_2O_4} + P_{NO_2} = P_D + P_M$$

$$P^2_M + K \cdot P_M - K \cdot P_T = 0$$

$$P_M = \frac{-K + \sqrt{K^2 + 4 \cdot K \cdot P_T}}{2}$$

$$P_D = P_T - P_M$$

Worksheet 15.3 computes the partial pressures of the two gases in atmospheres when the total pressure of the system is entered in cell B3. The numerical value of the equilibrium constant is expressed in atmospheres, and other temperatures may be entered in cell B2. Formula List 15.3 shows the cells that are defined and the formulas that are entered in the spreadsheet.

	A	B	C	D
1	N2O4 =	2 NO2	Equilibrium	
2	**Kp =**	**6.49**	100 C	
3	**P (total) =**	**2.00**		
4				
5		Equilibrium		
6	P (N2O4) =	0.396		
7	P (NO2) =	1.604		
8			**6.49**	**= Kp**
9	P (total) =	**2.000**		

Worksheet 15.3 Effect of Total Pressure

Formula List 15.3 Formulas for Worksheet 15.3

Cell $B$2	Defined as K
Cell $B$3	Defined as PT
Cell $B$7	Defined as PM
Cell B6	=PT-PM
Cell B7	=(-K+(K^2+4*K*PT)^0.5)/2

Table 15.1 presents the equilibrium partial pressures of N_2O_4 and NO_2 when the external pressure is varied. At low pressure the reaction proceeds toward the side with the larger number of gas molecules, establishing an equilibrium position with relatively high partial pressures of NO_2 in comparison with N_2O_4. The reverse is true at high pressure.

Table 15.1 Partial Pressure of N_2O_4 and NO_2

P_T	$P_{N_2O_4}$	P_{NO_2}
0.1	0.001	0.099
1.0	0.119	0.881
10	4.56	5.44
100	77.6	22.4

The Effect of Volume

When a system of gases at equilibrium is subjected to a change in the total available volume, the position of equilibrium shifts as it does when the external pressure is changed. A decrease in volume has exactly the same effect on the equilibrium position as an increase in the external pressure. When considering a change in the total pressure, you should question if this change is the result of a reduction in the available volume. A change in the volume of the system produces a corresponding change in the concentrations of all gas species. When the change is the result of a volume reduction (an increase in pressure), the equilibrium position shifts toward that side of the balanced equation that has the fewest gas molecules. The addition of an inert gas has no effect on the position of equilibrium when all of the gases behave in an ideal manner. Worksheet 15.4 calculates the initial and equilibrium partial pressures for the dinitrogen tetraoxide-nitrogen dioxide system when the following are specified: the temperature, the volume, the initial moles of reactant, and the initial moles of product. Ideal gas behavior is assumed for both the reactant and the product.

	A	B	C	D
1	N2O4 =	2 NO2	Equilibrium	
2	0.119	0.881	<- moles	
3		30.6	<- volume	
4	0.08205	373.1	<- T (K)	
5				
6	**Kp =**	**6.49**	="100 C"	
7				
8		Initial	Equilibrium	
9	P (N2O4) =	=nD*Rg*T/V	=D-(-(4*M+K)+((4*M+K)^2-16*(M^2-K*D))^0.5)/8	
10	P (NO2) =	=nM*Rg*T/V	=M+(-(4*M+K)+((4*M+K)^2-16*(M^2-K*D))^0.5)/4	
11	P (total) =	=B9+B10	=C9+C10	
12				
13	**Q =**	**=B10^2/B9**	**=C10^2/C9**	**= Kp**
14			=IF(Q=K,"equilibrium"," ")	
15			=IF(Q>K,"Q>K shift toward reactants"," ")	
16			=IF(Q<K,"Q<K shift toward products"," ")	

Worksheet 15.4 Effect of Volume

LeChatelier's Principle predicts the correct pattern of behavior for any system that involves gaseous species. A reduction in the volume, which corresponds to an increase in the total pressure, shifts the equilibrium position to the side of the balanced equation that contains the fewest number of gas molecules. Formula List 15.4 lists the defined cells and formulas.

Formula List 15.4 Formulas for Worksheet 15.4

Cell $A$2	Defined as nD
Cell $A$4	Defined as Rg
Cell $B$2	Defined as nM
Cell $B$3	Defined as V
Cell $B$4	Defined as T
Cell $B$6	Defined as K
Cell $B$9	Defined as D
Cell $B$10	Defined as M
Cell $B$13	Defined as Q

The Effect of Temperature

The equilibrium position or the extent of reaction is altered by all of the following: a change in the temperature, a change in the concentration of one or more of the chemical species, and a change in the volume or pressure (with reactions involving gases where the number of gaseous molecules of reactants is not equal to the number of gaseous molecules of products). The equilibrium constant is altered by only a temperature change, and this effect can be predicted by LeChatelier's Principle. Exothermic reactions will favor the production of reactants at high temperature and the production of products at low temperature. The quantitative results of this effect can be calculated by the Gibbs-Helmholtz equation, which is developed in courses such as physical chemistry. We will be content to use the end result without worrying about the thermodynamic principles needed to develop this equation. The equation is very often expressed in the natural log form, but the exponential form is convenient for spreadsheet calculations.

$$K_2 = \mathrm{K}_1 \cdot \mathrm{e}^{H}$$

$$H = \frac{\Delta H^\circ}{\mathrm{R}}\left(\frac{1}{T_1} - \frac{1}{T_2}\right)$$

The equilibrium reaction between dinitrogen tetraoxide and nitrogen dioxide is exothermic. The value of ΔH° can be computed from any table of thermodynamic values (BLB Appendix). These calculations always use the convention of taking the sum of the enthalpies of the products minus the sum of the enthalpies of the reactants. Most tables present values for 25°C for this reaction:

N_2O_4 (*g*) ΔH^o_f = 9.66 kJ/mol

NO_2 (*g*) ΔH^o_f = 33.84 kJ/mol

$\Delta H°$ = 58.02 kJ

Worksheet 15.5 calculates the equilibrium constant for any temperature based on this calculated value of $\Delta H°$ at 25°C. In truth, $\Delta H°$ is a function of temperature, but it is relatively constant over a restricted temperature interval. Recalling the equilibrium values previously computed for this reaction at a temperature of 100°C shows that the equilibrium position favors the concentration of the reactants relative to that of the products at the lower temperature because of the endothermic character of this reaction. For this spreadsheet, the temperature and equilibrium constant are 100°C and 0.212. Altering the values of the temperature in cell B4 will show the magnitude of the changes in the equilibrium constant and the equilibrium position relative to the chosen reference temperature. If an experimental value for the equilibrium constant at another temperature is known, that value can be entered in cell B7 along with the necessary change in cell C7.

	A	B	C	D
1	N2O4 =	2 NO2	Equilibrium	
2	**0.212**	**0.212**	**<- moles**	
3		**1**	**<- volume**	
4		**323.15**	**<- T(K)**	
5		**58.02**	**<- delta H(kJ)**	
6				
7	Keq =	0.212	100	<- t (C)
8	**Keq =**	**0.0117**	50.0	<- t (C)
9				
10		Initial	Equilibrium	
11	[N2O4] =	0.212	0.289	M
12	[NO2] =	0.212	0.058	M
13				
14	**Q =**	**0.212**	**0.0117**	**= Keq**
15				
16			Q>K shift toward reactants	

Worksheet 15.5 Effect of Temperature

Worksheet 15.5F illustrates the important formulas that are used in this spreadsheet. Formula List 15.5 lists the defined cells formulas.

	B
8	=Kr*EXP(dH*(1/(Tr+273.15)-1/T)/0.0083145)
9	
10	Initial
11	=nD/V
12	=nM/V
13	
14	**=B12^2/B11**

	C
11	=D-(-(4*M+K)+((4*M+K)^2-16*(M^2-K*D))^0.5)/8
12	=M+(-(4*M+K)+((4*M+K)^2-16*(M^2-K*D))^0.5)/4
13	
14	**=C12^2/C11**
15	=IF(Q=K,"equilibrium"," ")
16	=IF(Q>K,"Q>K shift toward reactants"," ")
17	=IF(Q<K,"Q<K shift toward products"," ")

Worksheet 15.5F Formulas for Worksheet 15.5

Formula List 15.5 Formulas for Worksheet 15.5

Cell $A$2	Defined as nD
Cell $A$4	Defined as Rg
Cell $B$2	Defined as nM
Cell $B$3	Defined as V
Cell $B$4	Defined as T
Cell $B$5	Defined as d
Cell $B$7	Defined as Kr
Cell $B$8	Defined as K
Cell $B$11	Defined as D
Cell $B$12	Defined as M
Cell $B$14	Defined as Q
Cell $C$7	Defined as Tr

Equilibrium Concentrations by Approximation Methods

The mathematical treatment of many chemical systems at equilibrium produces polynomial equations of third degree or higher. In these cases, the chemist must find the single real root that has chemical significance in order to obtain the equilibrium concentrations. The initial concentrations of the chemical species serve to identify the root that has meaning with respect to the chemistry. The technique for computing the root of chemical significance has to focus on the correct root and ignore all of the other possible roots that will produce negative concentrations when the system is at equilibrium. The technique that we will use for obtaining this root is known as the Newton-Raphson method. The graphical interpretation of this method is represented by a two-dimensional plot of the function. The roots are represented by those points where the function crosses the x-axis. The numerical values of these roots are such that the function when evaluated at these points will have a value of zero. This technique is an iterative process where an initial value, x_0, is specified and used to calculate the value of the function and the slope of the function at that point. From these values a new value, x_1, is calculated from the point at which the tangent (the slope of the function) crosses the x-axis. The new value, x_1, is then used to calculate the value of the function and the slope at x_1, which serves to produce another value, x_2. This iterative process continues until conditions that are specified by the user are achieved. The development of the polynomial function that describes the chemical system is identical to the process that has been used before in this chapter. The development of the equation or function that describes the slope requires a process for obtaining the derivative of the function. This is a common technique in calculus, and for our purposes the process is easily accomplished because the chemical systems of interest are described by polynomials. The only terms of concern to us involve a constant multiplied by x raised to a power. The standard form for this process can be found in any calculus text or handbook of mathematics. The derivative of this polynomial term

$$f(x) = C \cdot x^n$$

is

$$f'(x) = C \cdot n \cdot x^{n-1}$$

The dinitrogen tetraoxide-nitrogen dioxide system will serve as the first example of the application of this approximation technique.

$$N_2O_4\,(g) = 2\,NO_2\,(g)$$

Equilibrium	D - x	M + 2x

$$K_{eq} = \frac{[NO_2]^2}{[N_2O_4]} = \frac{(M + 2x)^2}{D - x}$$

$$f(x) = 4x^2 + (4 \cdot M + K)x + M^2 - K \cdot D$$

$$f'(x) = 8x + 4 \cdot M + K$$

Worksheet 15.6 provides a strategy for investigating the acquisition of the chemically meaningful root of the second degree polynomial function that describes this system. The initial guess is a critical factor for success.

	A	B	C	D	E
1	N2O4 =	2 NO2	Equilibrium		
2	**Keq =**	**0.212**	100 C		
3					
4		Initial	Equilibrium		
5	[N2O4] =	0.212	0.212	root 1 =	0
6	[NO2] =	0.212	0.212	root 2 =	-0.265
7					
8	**Q =**	**0.212**	**0.212**	**= Keq**	
9			equilibrium		
10					
11					
12					
13	**Newton - Raphson Iteration Technique**				
14	f(x)	0	0	<-guess	
15	f'(x)	1.06			
16	x	1.6666E-18			
17		Initial	Equilibrium		
18	[N2O4] =	0.212	0.212		
19	[NO2] =	0.212	0.212		
20					
21		Keq recalc =	0.212		

Worksheet 15.6 Newton-Raphson Iteration Technique

This spreadsheet computes the equilibrium concentrations as before from the quadratic solution of the second degree equation. In addition, both roots of this function are calculated by the quadratic solution. The bottom half of the spreadsheet uses the Newton-Raphson technique for finding roots to compute the value of the single root of chemical significance. The value of the function and the value of the derivative of the function are computed by this iterative method. The value of x is computed, and from this, the values of the equilibrium concentrations. The guess value, x_0, is computed for use when and if the equilibrium concentrations become negative. If, after the first 100 iterations, the equilibrium concentration of one or more species is negative, use the cursor to select cell B16, which contains the value of x. Then click the cursor on the formula bar (use care so that you do not disturb the formula that is shown), and then select and click on the square that contains the check. This will initiate a new series of iterations in search of the chemically significant root. The guess value, x_0, will be used for the initial calculation during this series of iterations. The chance of computing an incorrect solution is always a possibility. When this has happened, the recalculated equilibrium constant in conjunction with the equilibrium concentrations serve to alert the user.

Worksheet 15.6F shows the formulas used to find the root of chemical significance. If the iteration procedure converges on a root that results in negative equilibrium concentrations, the conditional statement in cell B16 initiates a new series of iterations with the first value of x being equal to the value in cell C14. The intent of this process is the acquisition of the real root that has chemical meaning. The formula in cell C14 is derived from the original polynomial function by ignoring all of the terms with a degree higher than one. The value computed from this approximation will vary significantly from the correct value of x, but it is close enough to allow the iteration procedure to converge on the root of chemical significance. If negative values for the equilibrium concentrations appear, select the value of x, cell B16, and initiate a new series of iterations as described in the previous paragraph. Both roots of the polynomial are also evaluated.

	B	C
5	0.212	=D-(-(4*M+K)+((4*M+K)^2-16*(M^2-K*D))^0.5)/8
6	0.212	=M+(-(4*M+K)+((4*M+K)^2-16*(M^2-K*D))^0.5)/4
7		
8	**=B6^2/B5**	**=C6^2/C5**
9		=IF(Q=K,"equilibrium"," ")
10		=IF(Q>K,"Q>K shift toward reactants"," ")
11		=IF(Q<K,"Q<K shift toward products"," ")
12		
13	**Raphson Iteration**	
14	=4*x^2+(4*M+K)*x+M^2-K*D	=(K*D-M^2)/(4*M+K)
15	=8*x+4*M+K	
16	=IF(C18<0,X0,IF(C19<0,X0,x-Y/DY))	
17	Initial	Equilibrium
18	=B5	=D-x
19	=B6	=M+2*x
20		
21	Keq recalc =	=C19^2/C18

	D	E
5	root 1 =	=(-(4*M+K)+((4*M+K)^2-16*(M^2-K*D))^0.5)/8
6	root 2 =	=(-(4*M+K)-((4*M+K)^2-16*(M^2-K*D))^0.5)/8

Worksheet 15.6F Newton-Raphson Technique Formulas

For these iterations, the default parameters of Microsoft® Excel Student Edition version 2.1s are used. First, the Options choice is selected from the menu bar. Then, Calculation is chosen from the Options menu. The dialog boxes that are presented have a choice for the Maximum Iterations (default 100) and the Maximum Change (default 0.001). Before closing this window, select and click on the box labeled Iteration. Setting this allows the spreadsheet to compute algorithms that have circular logic. The cells that are defined with specific symbols are listed in Formula List 15.6.

Formula List 15.6 Formulas for Worksheet 15.6

Cell $B$2	Defined as K
Cell $B$5	Defined as D
Cell $B$6	Defined as M
Cell $B$8	Defined as Q
Cell $B$14	Defined as Y
Cell $B$15	Defined as DY
Cell $B$16	Defined as x
Cell $C$14	Defined as X0

The Newton-Raphson method for finding roots of a function and the spreadsheet techniques for finding the single real root of chemical significance developed in the preceding paragraphs are applicable to more complex chemical systems. The Haber reaction is an intriguing chemical system because the mathematical analysis produces a fourth degree polynomial that must be solved. From the mathematical treatment of the chemical equilibrium there exists one real root that produces positive concentrations for all of the species, and that simultaneously obeys the constraints imposed by the initial concentrations and the conservation of mass. The Haber reaction describes the production of ammonia from chemical reaction of nitrogen and hydrogen.

$$N_2\,(g) \ + \ 3\,H_2\,(g) \ = \ 2\,NH_3\,(g)$$

Equilibrium	$N - x$	$H - 3x$	$A + 2x$

$$K_{eq} = \frac{[NH_3]^2}{[N_2]\cdot[H_2]^3} = \frac{(A + 2x)^2}{(N - x)\cdot(H - 3x)^3}$$

$$f(x) = 27Kx^4 - 27K(N+H)x^3 + (9H\cdot K(3N+H)-4)x^2 + (K(9N+H)H^2+4A)x + K\cdot N\cdot H^3 - A^2$$

$$f'(x) = 108Kx^3 - 81K(N+H)x^2 + (18H\cdot K(3N+H)-8)x + K(9N+H)H^2 + 4A$$

Worksheet 15.7 computes the equilibrium partial pressures of nitrogen, hydrogen, and ammonia from the initial partial pressures of those molecules and the equilibrium constant. The guess value, x_0, is computed for use when the equilibrium concentrations become negative. If the algorithm converges on an inappropriate root, producing negative equilibrium concentrations, use the cursor to select cell B15, which contains the formula for computing the value of x. Activate the formula bar by clicking the cursor on it, and then click on the square that contains the check. This initiates a new series of iterations. The guess value, x_0, will be used for the first calculation during this series of iterations. The recalculated equilibrium constant in conjunction with the equilibrium concentrations serve to alert the user when an incorrect solution has been computed.

	A	B	C	D
1	N2 + 3 H2 =	2 NH3	Equilibrium	
2	**Kp =**	**0.0000451**	450 C	
3				
4		Initial		
5	P (N2) =	10		
6	P (H2) =	10		
7	P (NH3) =	10	**0.01**	**= Q**
8				
9			Q>K shift toward reactants	
10				
11				
12	**Newton - Raphson Iteration Technique**			
13	f(x)	0	-2.460977479	<-guess
14	f'(x)	-12.792034		
15	x	-3.7732639		
16		Initial	Equilibrium	
17	P (N2) =	10	13.773	
18	P (H2) =	10	21.320	
19	P (NH3) =	10	2.453	
20				
21			**4.51E-05**	**= Keq**

Worksheet 15.7 The Haber Reaction

Worksheet 15.7F shows the formulas used for acquiring the chemically significant root. The initial value, x_0, is calculated by the formula in cell C13, which is derived from the original polynomial function by ignoring all of the terms with a degree higher than one. This value is an approximation that allows the iteration procedure to converge on the single root of chemical significance. The cells that are defined with specific symbols are presented in Formula List 15.6. For chemical systems that generate polynomial functions that have two or more roots, it is a reasonable expectation that there is only one set of conditions that represent the equilibrium state. The partial pressures and concentrations values must be positive in all cases.

	B
13	=27*K*x^4-27*K*(N+H)*x^3+(9*H*K*(3*N+H)-4)*x^2-(K*(9*N+H)*H^2+4*A)*x+K*N*H^3-A^2
14	=108*K*x^3-81*K*(N+H)*x^2+(18*H*K*(3*N+H)-8)*x-K*(9*N+H)*H^2-4*A
15	=IF(N<0,X0,IF(H<0,X0,IF(A<0,X0,x-Y/DY)))

	C
7	**=A^2/(N*H^3)**
8	=IF(Q=K,"equilibrium"," ")
9	=IF(Q>K,"Q>K shift toward reactants"," ")
10	=IF(Q<K,"Q<K shift toward products"," ")
11	
12	**Iteration**
13	=(K*N*H^3-A^2)/(K*(9*N+H)*H^2+4*A)
14	
15	
16	Equilibrium
17	=N-x
18	=H-3*x
19	=A+2*x
20	
21	**=C19^2/(C17*C18^3)**

Worksheet 15.7F Formulas for Worksheet 15.7

Formula List 15.7 Formulas for Worksheet 15.7

Cell $B$2	Defined as K
Cell $B$5	Defined as N
Cell $B$6	Defined as H
Cell $B$7	Defined as A
Cell $B$13	Defined as Y
Cell $B$14	Defined as DY
Cell $B$15	Defined as x
Cell $C$7	Defined as Q
Cell $C$13	Defined as X0

Problems

1. $Br_2\ (g) = 2\ Br\ (g)$ @1223 K, $K_p = 3.28 \times 10^{-3}$ atm

   Construct a spreadsheet that will calculate the equilibrium partial pressures and concentrations for both species at equilibrium. Calculate the partial pressure of the bromine atoms when bromine gas has an initial pressure of 2 atm. Enhance the spreadsheet so that volume and the initial moles of bromine may be entered as data.

2. $Br_2\,(g) \;=\; 2\,Br\,(g)$ @1223 K, $K_p = 3.28 \times 10^{-3}$ atm

   Construct a spreadsheet for computing the equilibrium partial pressures at different temperatures. For this equilibrium at 1200 K, use a value of $\Delta H° = 203$ kJ (at 25°C $\Delta H° = 193$ kJ). Compare your calculations with $K_p = 1.40 \times 10^{-3}$ atm @ 1173 K. Calculate the partial pressure of the bromine atoms when bromine gas has an initial pressure of 2 atm at a temperature of 1400 K.

3. $2\,HI\,(g) \;=\; H_2\,(g) + I_2\,(g)$ @698.6 K, $K_p = 1.83 \times 10^{-2}$

   Create a spreadsheet that will compute the equilibrium partial pressures of all species at equilibrium. Include options for entering the volume, moles of material, and temperature. For the temperature dependence, use $\Delta H° = 10.4$ kJ, which is valid at 25°C. Calculate the equilibrium partial pressures of all species when the initial partial pressure of hydrogen iodide is 2.00 atm. Repeat this calculation with the initial partial pressures of hydrogen and iodine at 1.00 atm each.

4. $I_2\,(g) + Br_2\,(g) \;=\; 2\,IBr\,(g)$ @150°C, $K_p = 2.80 \times 10^{2}$

   Create a spreadsheet for computing the equilibrium partial pressures of all species from data on the volume, moles of material, and temperature. Use $\Delta H° = -11.4$ kJ at 25°C. Calculate the equilibrium partial pressures of all species when the initial partial pressures of both iodine and bromine are 5.00 atm each.

5. $2\,NO\,(g) + O_2\,(g) \;=\; 2\,NO_2\,(g)$ @184°C, $K_p = 1.48 \times 10^{4}$ atm^{-1}

   Design and create a spreadsheet for computing the equilibrium partial pressures of all species from data on the initial partial pressures by the Newton-Raphson technique. Use $\Delta H° = -113$ kJ at 25°C for the temperature dependence. Calculate the equilibrium partial pressures of all species when the initial partial pressures of oxygen and nitric oxide are 0.210 atm and 0.00050 atm. Calculate the equilibrium partial pressures of all species when nitric oxide and oxygen are initially present at 0.0500 atm and a temperature of 100 °C.

6. $2\,CO_2\,(g) \;=\; 2\,CO\,(g) + O_2\,(g)$ @3000 K, $K_p = 0.805$ atm

   Create a spreadsheet for computing by the Newton-Raphson technique the equilibrium partial pressures from the initial partial pressures. Use $\Delta H° = 566$ kJ at 25°C. Calculate the equilibrium partial pressures when the initial partial pressures of oxygen and carbon dioxide are 0.21 atm and 0.030 atm. Calculate the equilibrium partial pressures when carbon dioxide is the only gas initially present at 50.0 atm.

7. $N_2\,(g) + 3\,H_2\,(g) = 2\,NH_3\,(g)$ @300°C, $K_p = 4.34 \times 10^{-3}$ atm^{-2}

   Construct a spreadsheet for computing the equilibrium partial pressures from the initial partial pressures by the Newton-Raphson technique. Use $\Delta H^\circ = -105$ kJ. Calculate the equilibrium partial pressures when the initial partial pressures of all of the gases are 10.0 atm and the temperature is 300°C. Compare the computed equilibrium constant to $K_p = 4.51 \times 10^{-5}$ atm^{-2} at 450°C. Repeat the equilibrium calculations for all gases initially present at 10.0 atm and 450°C.

8. $N_2\,(g) + 3\,H_2\,(g) = 2\,NH_3\,(g)$ @450°C, $K_p = 4.51 \times 10^{-5}$ atm^{-2}

   Construct a spreadsheet for computing by the Newton-Raphson technique the equilibrium partial pressures from the total pressure of the system. Use $\Delta H^\circ = -105$ kJ. Calculate the equilibrium partial pressures when the total pressure is 1000 atm and 450°C. Calculate the equilibrium partial pressures when the total pressure is 10.0 atm at 450°C. Repeat the calculations at 1000 atm and 10.0 atm for a temperature of 300°C.

9. $2\,NOBr\,(g) = 2\,NO\,(g) + Br_2\,(g)$ @373°C, $K_p = 0.416$ atm

   Create a spreadsheet for computing the equilibrium partial pressures from data on the initial partial pressures by the Newton-Raphson method. Use $\Delta H^\circ = 47.8$ kJ at 25°C for the temperature dependence.

10. $2\,NOCl\,(g) = 2\,NO\,(g) + Cl_2\,(g)$

    Determine Kp and ΔH° from the following data. The total pressure is at equilibrium. Individual pressures are the initial partial pressures. Pressures are in mm of Hg.

t (°C)	P_{eq}	P_{NOCl}	P_{NO}	P_{Cl_2}
230	539.0	208.7	0	312.4
308	648.4	241.0	0	360.6
399	781.2	278.5	0	416.9
465	1027.7	589.6	218.2	0

16

Acid-Base Equilibria

The acidic or basic properties of a solution are determined by the relative amounts of hydrogen ion and hydroxide ion. A series of spreadsheet techniques are developed for calculating $[H^+]$ and the corresponding *pH* for weak acids or weak bases, with or without salts that contain conjugate ions, and for salts with and without the associated conjugate ions. The traditional methods for calculating $[H^+]$ have relied on approximate numerical solutions for chemical systems that were incompletely specified. We will first describe improved numerical techniques (a reduction in the need for ignoring functional terms that are significant) for obtaining mathematical solutions for chemical systems that are incompletely specified, and then demonstrate the ease with which valid solutions for completely described chemical systems may be obtained with any desired precision. These modern computational techniques highlight the uncertainty caused by experimental error. (BLB Chap. 16 and 17)

For aqueous solutions of weak acids that ignore the water equilibrium, we will compare three methods of calculating $[H^+]$: the usual approximation, the exact mathematical solution using the quadratic expression, and an iteration technique that demonstrates the power of Excel as a numerical analysis tool. Our next approach considers techniques for calculating the *pH* of weak acid buffers or weak base buffers with the provision that either a weak acid or weak base be initially present. As our last consideration of incompletely specified chemical systems, we will approach the mathematical solution from the other extreme of having a salt initially present with or without its conjugate species. The range over which these calculations are mathematically valid will be considered.

As a final consideration, we will examine the mathematical solution that may be obtained for the completely specified aqueous acid-base equilibria. The systems we will explore are a weak acid and a salt with the conjugate anion of the weak acid plus the water dissociation equilibrium, and the complementary case of a weak base and a salt with the conjugate cation of the weak base plus the water dissociation equilibrium. The mathematical solution of these chemical equilibria generates a cubic equation that may be solved by the Newton-Raphson method for the root that has chemical significance. The simplicity and accuracy of these mathematical solutions, in conjunction with the unlimited range over which these results are chemically and mathematically valid, make this the method of choice.

Weak Acids

The hydrogen ion concentration or the *pH* of an aqueous solution that contains only a weak acid is usually determined by a method that ignores the "real" concentration of the remaining molecular acid after the dissociation equilibrium has been achieved. A closed solution that considers the "real" concentration of the acid at equilibrium may be obtained from the quadratic equation. A third option is provided by the use of Microsoft® Excel spreadsheet software with the iteration option which allows the user to solve this type of problem without the necessity of manipulating the terms of the quadratic equation. These techniques in all cases are applied to an incompletely specified system as the water dissociation equilibrium is ignored. The complete chemical system will be considered later in this chapter under the heading of The Cubic Solution For Aqueous Acid-Base Equilibria.

For a weak acid, the chemical equilibrium is represented by the following equation with the associated acid-dissociation constant.

$$HX = H^+ + X^- \qquad K_a = \text{a numerical constant}$$

The mathematical expression for the hydrogen ion concentration is

$$[H^+]^2 = K_a \cdot (C_a - [H^+])$$

Worksheet 16.1 illustrates the use of a spreadsheet for the calculation of $[H^+]$ and pH by three different methods: first, the solution by the usual approximation method, which ignores the subtraction of $[H^+]$ from C_a, then, an exact mathematical solution that is produced by selecting only the positive root of the quadratic equation that results from the polynomial form of the equilibrium expression, and last, an iteration method that is available with Excel. This latter technique is very convenient, although there are pitfalls because the solution requires the use of the square root function which can produce an undefined value (the square root of a negative number). The solution of the quadratic equation and the iteration method produce results that are exact mathematical solutions (within the limits dictated by the significant digits) for this chemical system, which includes only the dissociation of the weak acid.

	A	B	C
1	Hydrofluoric Acid - Comparing Methods		
2			
3	6.80E-04	= Ka	
4	0.00168	= Initial [HF]	
5			
6		[H+]	pH
7	H+ approximate	1.07E-03	2.97
8	H+ quadratic	7.82E-04	3.11
9	H+ iteration	7.82E-04	3.11

Worksheet 16.1 Comparing Methods of Calculating [H+]

Worksheet 16.1F shows the formulas used in these computations. You should convince yourself that these functions are appropriate for the determination of [H+]. If you have difficulty with the iteration technique, you should study the next paragraph. Cell A3 is defined as K and cell A4 as IC. These cells contain the acid dissociation equilibrium constant and the initial or formal concentration of the weak acid.

	A	B	C
1	**Hydrofluoric Acid -Comparing Methods**		
2			
3	0.00068	="= Ka"	
4	0.00168	="= Initial [HF]"	
5			
6		[H+]	pH
7	H+ approximate	=(K*IC)^0.5	=-LOG10(B7)
8	H+ quadratic	=(-K+(K^2+4*K*IC)^0.5)/2	=-LOG10(B8)
9	H+ iteration	=(K*(IC-B9))^0.5	=-LOG10(B9)

Worksheet 16.1F Formulas For Worksheet 16.1

A simple technique is available with Excel that allows consideration of the [H+] term in the parentheses, and thus produces a calculated result that is superior to the usual approximation, which ignores the subtraction of the equilibrium value of [H+] from the initial concentration of the acid. Calculations of the hydrogen ion concentration for dilute acids that are moderately strong are only valid with the inclusion of this term. This point is illustrated by the computations for phosphorous acid that are illustrated later in this chapter. If you want to calculate [H+] in any cell, then enter

$$(K_a \cdot (C_a - [H^+]))^{.5}$$

in that cell. After the program responds, "Can't resolve circular references," you select Calculation from the Options menu and check Iteration. You may then select the number of iterations and the numerical change between iterations that will terminate the calculations. The use of the iteration option is sensitive to functions that require the square root as it is possible to generate a negative value which is undefined. This happens (labeled as #NUM!) when the value of the cell becomes very small. This may also happen on the first computation with a new spreadsheet. Excel uses zero as the initial value in cells that have not been computed but whose values are needed by other cells. This only happens during the first iteration because the successive computations use the current values of the cells. Starting a new iteration process with the current value can be used successfully to approach a discontinuity in a function. The chances of successfully calculating values for algorithms that have circular logic are increased by using the manual calculation option of the spreadsheet and by making all changes in the values before the calculations are commenced.

Worksheet 16.2 illustrates the values of [H+] produced by the three methods for a specific weak acid at several different initial concentrations.

	A	B	C	D
1		**Acetic Acid - Comparing Methods**		
2				
3	1.79E-05	=Ka		
4	1	= [Stock Soln]		
5				
6	Dilutions	approximate	quadratic	iteration
7		[H+]	[H+]	[H+]
8	1	4.23E-03	4.22E-03	4.22E-03
9	0.1	1.34E-03	1.33E-03	1.33E-03
10	0.01	4.23E-04	4.14E-04	4.14E-04
11	0.001	1.34E-04	1.25E-04	1.25E-04
12	0.0001	4.23E-05	3.43E-05	3.43E-05
13	0.00001	1.34E-05	7.15E-06	#NUM!

Worksheet 16.2 Increasing Dilutions of Acetic Acid

These results for acetic acid with $K_a = 1.79 \times 10^{-5}$ and with initial acid concentrations of 0.1, 0.01, ..., 1×10^{-5} M illuminate the differences. Calculation of [H+] for a solution that has an initial concentration of 1×10^{-5} M is not possible by the iteration method because the value becomes negative after a few iterations. The calculation for an initial concentration of 1×10^{-4} M requires approximately 400 iterations and patience in achieving initial values that are close to the final solution. Different weak acids will produce *pH* values for the approximate and quadratic solutions that diverge at different levels of initial concentration. Initial weak acid concentrations that produce a $pH > 6.5$ should not be taken at face value unless the water dissociation equilibrium has also been considered.

A modification of the previous dilution is illustrated as Worksheet 16.3 to provide an example of the increase in the percent ionization of the molecular species as the solution is diluted. For this calculation, the quadratic solution technique will be used to calculate the percent of the molecular acid ionized. The equation used to provide the numeric solutions is

$$(- K_a + (K_a^2 + 4 \cdot K_a \cdot F)^{.5})/(2 \cdot F)$$

where F is the formal or initial concentration of the acid and K_a is the acid equilibrium constant. With increasing dilution, weak acids initially undergo a dramatic increase in the extent of ionization that must ultimately slow as this percentage asymptotically approaches the limit of complete ionization. A curve of this nature must have an inflection point, as can be seen with the graphical results of the percent ionization for acetic acid that is shown in Chart 16.3 on the next page. Be careful with your interpretation of this graph, as the abscissa has a nonlinear scale.

	A	B
1	**Acetic Acid - Extent of Ionization**	
2	1.79E-05	=Ka
3	0.1	[Original Soln]
4		
5	Dilutions	quadratic
6		% Ionization
7	0.1	1%
8	0.01	4%
9	0.001	13%
10	0.0001	34%
11	0.00001	71%
12	0.000001	95%
13	0.0000001	99%

Worksheet 16.3 Extent of Ionization of Acetic Acid

The cells in column B contain formulas that are analogous to Cell B7.

```
=(-K+(K^2+4*K*A7)^0.5)/(2*A7)
```

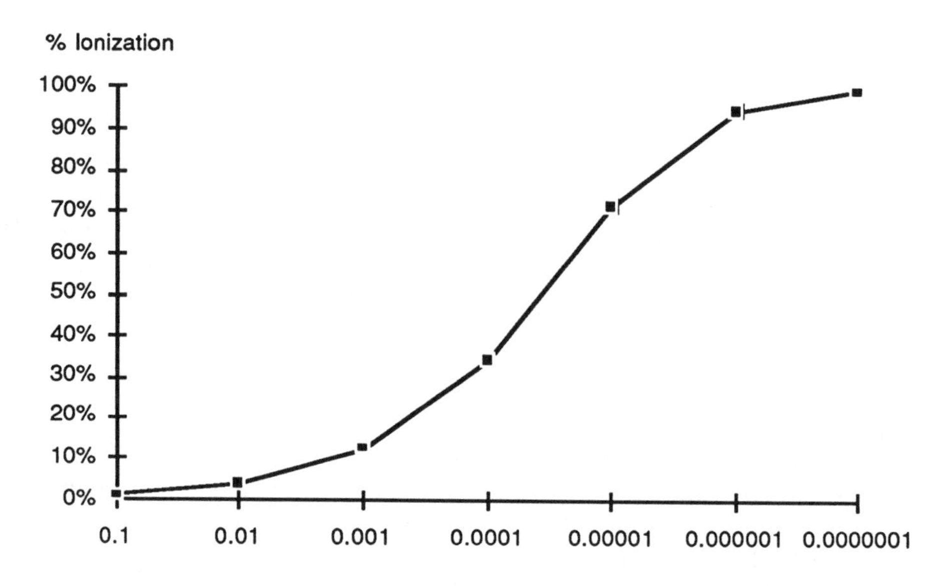

Chart 16.3 Ionization of Acetic Acid

Phosphorous acid, H_3PO_3, is a strong enough acid, $K_{a1} = 1.6 \times 10^{-2}$, to cause very early divergence in the results obtained by the approximation method as opposed to the quadratic solution. These results are illustrated as Worksheet 16.4 and Chart 16.4, where only the first acid dissociation reaction has been considered. When the phosphorous acid solutions have a concentration of 1.0×10^{-3} M or less, the different methods of calculating *pH* result in an error of 0.62 units or larger.

	A	B	C
1	Phosphorous Acid - Comparing Methods		
2	Ka =	1.60E-02	
3	Original Soln	0.1	
4			
5	Dilutions		
6		pH approx	pH quad
7	0.1	1.4	1.5
8	0.01	1.9	2.2
9	0.001	2.4	3.0
10	0.0001	2.9	4.0
11	0.00001	3.4	5.0
12	0.000001	3.9	6.0

Worksheet 16.4 Comparing Numeric Methods

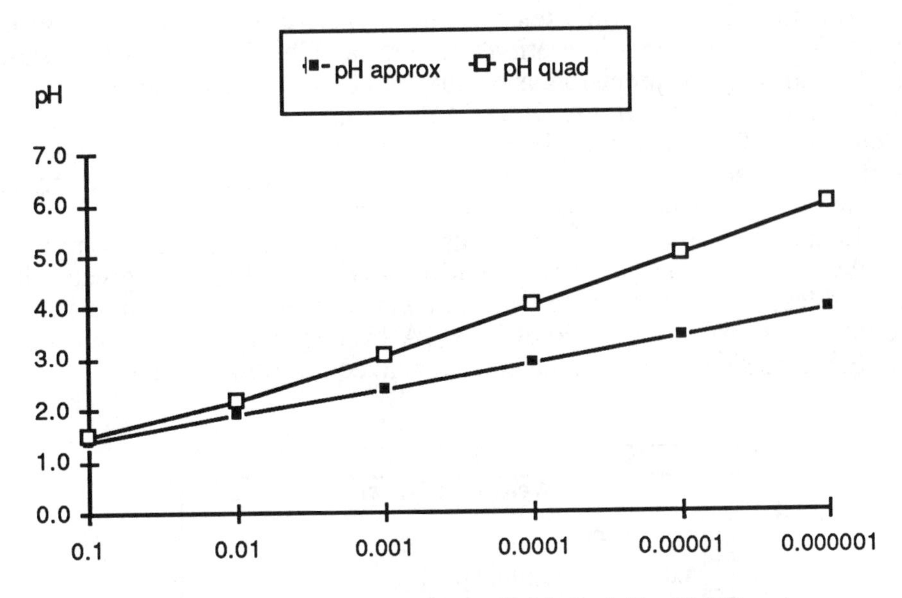

Chart 16.4 Comparing Numeric Methods For H_3PO_3

Weak Acid Buffers

When a solution contains significant concentrations of both the weak acid and the conjugate base of that acid, the solution is called a buffer because it will moderate the effect of adding either an acid or a base. Numerical calculations for the $[H^+]$ and pH may be achieved in a fashion similar to that of the previous section. The usual method is one of ignoring both the subtraction of $[H^+]$ from C_a and the addition of $[H^+]$ to C_b. An alternate method is furnished by selecting the positive root of the quadratic solution of the polynomial form of the equilibrium expression. A simple but accurate technique uses the iteration option that is available with Excel. With a buffer solution, it is necessary to solve the following equation.

$$[H^+] = K_a \cdot (C_a - [H^+]) / (C_b + [H^+])$$

Worksheet 16.5 computes the concentrations of all species, the percentage of the species in the non- and protonated form, and the pH of the solution at equilibrium. As an indication of the validity of the calculation, the last step in cell B11 recalculates the acid equilibrium constant from the computed concentrations. The iterative computation of $[H^+]$ takes place in cell A8.

```
=K*(CA-A8)/(CB+A8)
```

The initial interation uses a value of zero for A8 in the previous expression. Subsequent interations produce a value for $[H^+]$ that is within the desired range of precision as established by the value of the parameter, Maximum Change, in the Iteration option. If you initiate further calculations after a value has been calculated for cell A8, this value will be used as the starting point for the subsequent computation. The formulas used in this spreadsheet are presented in Worksheet 16.5F.

Equal initial concentrations of weak acid and its conjugate base produce the obvious result that $pH = pK_a$. Ratios of $[HA]/[A^-]$ that are greater than one and less than one produce values that are nonsymmetrical in relation to the pK_a value. When the ratio of $[HA]/[A^-]$ is larger than 1000, the result is unreliable, as shown by the recalculated acid equilibrium constant.

	A	B	C	D
1		**Weak Acid Buffer**		
2				
3	1.79E-05	**= Ka**		
4	0.01	**=Initial [HA]**		
5	0.1	**=Initial [A-]**		
6				
7	[H+]	[A-]	[HA]	[A-]
8	1.79E-06	1.00E-01	9.09%	90.91%
9				
10	[HA]	Ka(Recalc)	pH	pKa
11	1.00E-02	1.79E-05	**5.75**	4.75

Worksheet 16.5 Weak Acid Buffer

	A	B	C	D
1		**Weak Acid**	**Buffer**	
2				
3	0.0000179	**="= Ka"**		
4	0.01	**="=Initial [HA]"**		
5	0.1	**="=Initial [A-]"**		
6				
7	[H+]	[A-]	[HA]	[A-]
8	=K*(CA-A8)/(CB+A8)	=CB-A8	=(CA-A8)/(CA+CB)	=B8/(CA+CB)
9				
10	[HA]	="Ka(Recalc)"	pH	pKa
11	=CA-A8	=A8*B8/A11	**=-LOG10(A8)**	=-LOG10(K)

Worksheet 16.5F Formulas for Worksheet 16.5

A chemical system is most effective in buffering against a change in *pH* in either direction when the concentrations of the acid and its conjugate base are close to equal. If the ratio of $[HA]/[A^-]$ is greater than 1000, then the mathematical result produced by this technique is inaccurate. The next section presents an algorithm for an exact mathematical result. This is based on the solution of the quadratic expression for the dissociation of the weak acid or the weak base. This exact mathematical technique is limited by the absence of the equilibrium that is established with water.

Buffer Solutions

Chemical systems that resist a change in *pH* when a small amount of acid or base is added are called buffers. If a solution functions as a buffer, it must have both an acidic and basic species present. The most obvious buffers are solutions of a weak acid and its conjugate base or solutions of a weak base and its conjugate acid. The previous buffer equation

$$[H^+] = K_a \cdot (C_a - [H^+]) / (C_b + [H^+])$$

may be expressed in the form of a polynomial equation and mathematically solved for $[H^+]$. The solution of the quadratic equation produces two roots, one that will always yield a negative value for $[H^+]$ and one that will always yield a positive value (the value of $[H^+]$ may also be zero in this case if the initial concentration of the weak acid is zero). This result is easy to prove mathematically because the initial concentrations and the acid equilibrium constant are always positive. Thus, for a chemical system, the positive root is the desired expression. The generalized expression is

$$(-(CN + K) + ((CN + K)^2 + 4 \cdot AB \cdot K)^{.5})/2$$

where *CN* is the initial concentration of the conjugate species (base or acid) and *AB* is the initial concentration of either the weak acid or weak base.

This spreadsheet is illustrated as Worksheets 16.6A and 16.6B. The variables that the user supplies (such as K_a or K_b, and [HA] or $[HB^+]$) and the calculated values are changed in accordance with the text in cell B3 of the spreadsheet, which requires the user to enter either "acid" or "base". The critical formulas for this calculation are shown in Worksheet 16.6F.

This technique is valid for a wide range of initial concentrations for the weak acid or base and its conjugate. The calculated results are valid for either a weak acid or base as the initial concentration of the conjugate species approaches a limit of zero. The other extreme where the concentration of the weak acid or base approaches zero is not valid, and, in fact, the calculation is errored long before this limit. For weak acids, with $K_a > 10^{-6}$ and $[A^-]/[HA] < 1000$, the results calculated by this technique are valid. Decreasing values for K_a cause the calculations to be in error with a much smaller ratio of initial concentration of conjugate species to initial concentration of weak acid or base. As an example, the following conditions are the limit within which a valid result may be obtained, if $K_a = 1.05 \times 10^{-10}$ then $[A^-]/[HA] < 10$. The user may verify these results by using the technique for solving the cubic equation in the section titled Aqueous Acid-Base Equilibria.

	A	B	C
1	**Weak Acid or Base & Its Conjugate Species**		
2			
3	**acid or base:**	acid	<- You enter 1st
4	**Ka :**	1.79E-05	<- You enter
5	**[HA] :**	0.1	<- You enter
6	**[A-] :**	0.1	<- You enter
7		**OK**	
8	**[H+] :**	1.79E-05	
9	**[H+] :**	1.79E-05	
10	**pH :**	**4.75**	

Worksheet 16.6A A Weak Acid and Its Conjugate

	A	B	C
1	**Weak Acid or Base & Its Conjugate Species**		
2			
3	**acid or base:**	base	<- You enter 1st
4	**Kb :**	1.79E-05	<- You enter
5	**[B] :**	0.1	<- You enter
6	**[HB+] :**	0.1	<- You enter
7		**OK**	
8	**[OH-] :**	1.79E-05	
9	**[H+] :**	5.59E-10	
10	**pH :**	**9.25**	

Worksheet 16.6B A Weak Base and Its Conjugate

Worksheet 16.6A illustrates the weak acid, acetic, and Worksheet 16.6B the weak base, aqueous ammonia. The following cells are defined

$$[HA] = \$B\$5 = AB = [B]$$

$$[A^-] = \$B\$6 = CN = [HB^+]$$

	A	B
3	="acid or base:"	acid
4	=IF(B3="base","Kb :","Ka :")	0.0000179
5	=IF(B3="base","[B] :","[HA] :")	0.001
6	=IF(B3="base","[HB+] :","[A-] :")	0.1
7		=IF(CN/AB>100,"Wrong Solution Technique","OK")
8	=IF(B3="base","[OH-] :","[H+] :")	=(-(CN+K)+((CN+K)^2+4*AB*K)^0.5)/2
9	="[H+] :"	=IF(B3="base",0.00000000000001/B8,B8)
10	="pH :"	=-LOG10(B9)

Worksheet 16.6F Formulas for Worksheet 16.6

Acid-Base Properties of Salt Solutions

Salts may be considered to fall into one of four categories. These categories are based on the salt being produced hypothetically from an acid or a base that is weak or strong. We will consider the case of a salt of a strong acid and a weak base, and the converse, a salt of a strong base and a weak acid. This result is similar to that of the previous section, and it may be expressed in the form of a polynomial equation and solved for $[OH^-]$.

$$[OH^-] = K_a / K_w \cdot (C_b - [OH^-]) / (C_a + [OH^-])$$

As before, the solution of the quadratic equation produces two roots, one that will always yield a negative value for $[OH^-]$ and one that will always yield a positive value (the value of $[OH^-]$ may also be zero in this case if the initial concentration of the conjugate species, either the weak acid cation or the weak base anion, is zero). Thus, again for the chemical system, the positive root is the desired expression. The generalized expression is

$$(-(AB + K) + ((AB + K)^2 + 4 \cdot CN \cdot K)^{.5})/2$$

where CN is the initial concentration of the conjugate base, the anion, or the conjugate acid, the cation and AB is the initial concentration of either the protonated anion (the acid) or the cation minus a proton (the base). The equilibrium constant K is either K_w/K_a or K_w/K_b, as is appropriate. These results are a consequence of the relationship of conjugate species, which may be stated as a significant axiom of the Bronsted concept:

the stronger an acid, the weaker its conjugate base,
and the stronger a base, the weaker its conjugate acid.

$$K_w = K_a \cdot K_b$$

Worksheet 16.7 illustrates the computations for the salt of a weak acid or base. The formulas are presented in Worksheet 16.7F.

	A	B	C
1	**Salt of a Weak Acid or Base**		
2			
3	**acid or base:**	acid	<- You enter 1st
4	**Ka :**	1.79E-05	<- You enter
5	**[A-] :**	0.1	<- You enter
6	**[HA] :**	0	<- You enter
7		**OK**	
8	**Kb :**	5.59E-10	
9	**[H+] :**	7.47E-06	
10	**[H+] :**	1.34E-09	
11	**pH :**	**8.87**	

Worksheet 16.7 Salt of a Weak Acid

	A	B
3	="acid or base:"	acid
4	=IF(B3="base","Kb :","Ka :")	0.0000179
5	=IF(B3="base","[HB+] :","[A-] :")	0.1
6	=IF(B3="base","[B] :","[HA] :")	0.1
7		=IF(AB/CN>1,"Wrong Solution Technique","OK")
8	=IF(B6="base","Ka :","Kb :")	=0.00000000000001/K
9	=IF(B3="base","[OH-] :","[H+] :")	=(-(AB+KC)+((AB+KC)^2+4*CN*KC)^0.5)/2
10	="[H+] :"	=IF(B3="base",B9,0.00000000000001/B9)
11	="pH :"	=-LOG10(B10)

Worksheet 16.7F Formulas for Worksheet 16.7

Worksheet 16.7 computes the *pH* for the acetate anion which is the conjugate base, CN, of the weak acid, AB, acetic acid. This salt produces a solution that has a $pH > 7$. The mathematical solution calculated by this numeric method is valid if $[HA]/[A^-] < 1$ or if $[B]/[HB^+] < 1$. The calculated *pH* is valid in the limit of [HA] or [B] approaching and being equal to zero.

The Cubic Solution for Aqueous Acid-Base Equilibria

An aqueous solution of an acid, with or without the conjugate base, always involves not only the acid equilibrium but also the water dissociation equilibrium. This chemical system may be completely specified, and a mathematical solution for $[H^+]$ can be reached by solving the cubic equation that describes the behavior of this system. This cubic equation is developed by considering the three chemical equilibria that follow:

$$HX = H^+ + X^- \qquad K_a = \text{a numerical constant}$$

$$MX = M^+ + X^- \qquad \text{complete dissociation}$$

$$HOH = H^+ + OH^- \qquad K_w = 1.0 \times 10^{-14}$$

From these representations of the chemical equilibria, three algebraic equations may be developed. The first is based on a balance of the positive and negative electrical charge. The second is based on the equality of the total initial concentration of the anion, X^-, and the total equilibrium concentration of this anion which is in the protonated and non-protonated form. The last is based on the equilibrium expression.

$$[H^+] + [M^+] = [X^-] + [OH^-]$$

$$C_a + C_b = [HX] + [X^-]$$

$$K_a = [H^+][X^-] / [HX]$$

These three equations may be combined into one equation by the appropriate substitutions of equations one and two into equation three and the fact that $[M^+] = C_b$ and $[OH^-] = K_w/[H^+]$. The result is

$$K_a = \frac{[H^+](C_b + [H^+] - K_w/[H^+])}{C_a - [H^+] + K_w/[H^+]}$$

You should recognize some terms of this equation, as it is a more general solution to the problems that we have previously considered. When the condition $[H^+] >> K_w/[H^+]$ is valid, then this expression is identical to the equation that was used to solve the weak acid buffer problem. This inequality is correct when $[H^+] > 1 \times 10^{-7}$, and in practice the numerical results are such that the $K_w/[H^+]$ term can be ignored in both the numerator and denominator of this expression when $[H^+] > 1 \times 10^{-6}$. When this condition is valid and $C_b = 0$, this expression is identical to the equation for the dissociation of a weak acid.

An iterative program that utilizes the Newton-Raphson method of solving polynomial equations serves as the basic logic for this spreadsheet. This numerical technique requires evaluation of the cubic function at a specified concentration of $[H^+]$ and the first derivative at this point. The method for determining the first derivative is addressed in Appendix G.

The values of the function and its derivative in conjunction with the old value of [H+] are then used to calculate a new value of [H+]. The equations follow:

$$Y = H^3 + (K_a + C_b) \cdot H^2 - (K_w + K_a \cdot C_a) \cdot H - K_a \cdot K_w$$

$$DY = 3 \cdot H^2 + 2 \cdot (K_a + C_b) \cdot H - K_w - K_a \cdot C_a$$

$$H_{new} = H_{old} - Y/DY$$

Worksheet 16.8 illustrates the design of a spreadsheet that will compute the concentrations of all of the species under all conditions. This spreadsheet will compute valid concentrations for either acids or bases within the range $1 \times 10^{-19} < K_a$ or $K_b < 1 \times 10^5$.

	A	B	C
1	**Acid/Base-Water Equilibria**		
2			
3	**acid or base**	acid	**<-Enter 1st**
4	**Ka :**	1.80E-05	**<-Enter**
5	**[HA]init :**	0.000001	**<-Enter**
6	**[A-]init :**	0	**<-Enter**
7			
8	1.00E-14	Newton-Raphson	
9	**f(x)**	5.77779E-34	Quadratic
10	**f'(x)**	1.93063E-11	Solution
11	**[H+]eq :**	9.598E-07	9.499E-07
12	**[H+]eq :**	**9.598E-07**	9.499E-07
13	**[A-]eq :**	9.494E-07	9.499E-07
14	**[HA]eq :**	5.062E-08	5.013E-08
15			
16	**pH :**	**6.018**	6.022

Worksheet 16.8 Acid/Base-Water Equilibria

The Newton-Raphson technique for evaluating roots starts with an initial value of $[H^+]_1$ and computes subsequent values of $[H^+]_n$ to any specified precision. Subsequent iterations will converge on the single root that has chemical significance, as this root must be a positive real number greater than zero for this chemical system. By Descartes' rule, the cubic function has only one positive real root, so the numerical value of $[H^+]_n$ is determined unambiguously and expressed with as many significant digits as the experimental measurements of the equilibrium constants and initial concentrations warrant. Descartes' rule states that the number of positive real roots of a function with real coefficients is either equal to the number of variations in sign of the function (term by term) or less than that number by a positive even integer. It can be shown that this cubic function and the conditions dictated by the chemical system result in three real and unequal roots. Of these three real roots, one is positive and the other two are negative. The two real negative roots have no significance for the chemical system. Worksheet 16.8F shows the formulas used in this spreadsheet. The logic needed for the Newton-Raphson numerical technique is contained in cells B9 through B11. The conditional statements in cells B12, A4 to A6, A11, A12, and A14 are necessary for the dual approach, which allows computation from either the perspective of an aqueous acid or base.

	A
4	=IF(B3="base","Kb :","Ka :")
5	=IF(B3="base","[B]init :","[HA]init :")
6	=IF(B3="base","[HB+]init :","[A-]init :")
7	
8	0.00000000000001
9	f(x)
10	f'(x)
11	=IF(B3="base","[OH-]eq :","[H+]eq :")
12	[H+]eq :
13	=IF(B3="base","[HB+]eq :","[A-]eq :")
14	=IF(B12="base","[B]eq :","[HA]eq :")

	B
9	=B11^3+(K+CN)*B11^2-(KW+K*AB)*B11-K*KW
10	=3*B11^2+2*(K+CN)*B11-KW-K*AB
11	=IF(B11>0,B11-Y/DY,IF(QS<KW,0.0000001,QS))
12	=IF(B3="base",0.00000000000001/B11,B11)
13	=K*(AB+CN)/(B11+K)
14	=B11*(AB+CN)/(B11+K)
15	
16	=-LOG10(B12)

	C
11	=0.5*(-(K+CN)+((K+CN)^2+4*K*AB)^0.5)
12	=IF(B3="base",0.00000000000001/C11,C11)
13	=K*(AB+CN)/(C11+K)
14	=C11*(AB+CN)/(C11+K)
15	
16	=-LOG10(C12)

Worksheet 16.8F Formulas for Worksheet 16.8

Cells B13 and B14 compute the equilibrium concentrations of [A^-] and [HA]. These equilibrium concentrations are uniquely determined by [H^+]

$$[A^-] = \frac{K_a \cdot (C_a + C_b)}{[H^+] + K_a}$$

$$[HA] = \frac{[H^+] \cdot (C_a + C_b)}{[H^+] + K_a}$$

Cells B4 to B6 are defined as K, AB, and CN, where K is either K_a or K_b, AB is either C_a or C_b, and CN is then C_b or C_a (the conjugate species or the salt). Cell A8 is defined as KW, and cells B9 and B10 are defined as Y and DY, where Y is the cubic function and DY is the derivative of the cubic function. The mathematical solution of the quadratic expression that describes only the acid or base dissociation and neglects the water equilibrium is computed in cell C11. For the quadratic solution, the corresponding concentrations for the molecular acid or base and conjugate species are computed in cells C13 and C14. The conditional statement in cell B11 is necessary to ensure that the Newton-Raphson technique converges on the single positive real root. When the iteration produces a positive value for the hydrogen ion concentration, the new value is computed from the old value minus the quotient of the function divided by the derivative. When the iteration produces a negative value, there are two possible consequences. If the quadratic solution, QS, is less than K_w, a value of 1×10^{-7} is substituted for the current value of the hydrogen ion concentration. If the quadratic solution is equal to or greater than K_w, then the quadratic solution, QS, is substituted for the current value of the hydrogen ion concentration. These new initial values for [H^+] are necessary because a negative real root is very close to a value of zero when $K_a < 10^{-11}$ and $1 \times 10^{-4} < C_b < 0.1$ M. In fact, under these conditions, the mathematical solution of the quadratic approximation is zero. In order to ensure that the Newton-Raphson method does not converge on this negative root under these conditions, the nested conditional statement in cell B11 is necessary. This nested conditional statement resets the iteration value of [H^+] when the iterative technique produces a negative value.

This spreadsheet computes the *pH* for many more cases than first appears to be possible. You will be able to calculate the *pH* for all of the values listed in Table 16.1. The use of this method for calculating the *pH* of an aqueous acid-base solution is simple, quick, and as accurate as the data on the acid or base equilibrium constant and the initial concentrations of the species. This spreadsheet and the associated numerical techniques are valid for either weak or strong, acids or bases. Compare your computations by approximation methods with values produced by this technique.

Table 16.1 pH for Aqueous Acid-Base Equilibria

K_a or K_b	[HA] or [B]	[A-] or [BH+]	*pH*
1.80 x 10^{-5}	1.0 x 10^{-2}	0	3.382
	1.0 x 10^{-4}	0	4.464
	1.0 x 10^{-6}	0	6.018
	1.0 x 10^{-2}	1.0 x 10^{-2}	4.746
	0	1.0 x 10^{-2}	8.373
	0	0	7.000
2.30 x 10^{-11}	1.80 x 10^{-4}	0	6.955
	1.80 x 10^{-2}	1.80 x 10^{-2}	10.60
	0	1.80 x 10^{-2}	11.27
1.0 x 10^{5}	1.0 x 10^{-2}	0	2.00
	1.0 x 10^{-2}	1.0 x 10^{-2}	2.00
1.80 x 10^{-5}	1.0 x 10^{-2}	0	7.00
	1.0 x 10^{-2}	1.0 x 10^{-2}	12.00

Problems

1. Calculate the *pH* of a 2.47 x 10^{-3} M solution where:
   a) K_a = 6.8 x 10^{-4}
   b) K_a = 1.3 x 10^{-10}

2. What is the *pH* of a 0.0513 M benzoic acid and 1.26 x 10^{-3} M sodium benzoate solution? K_a = 6.28 x 10^{-5}

3. What is the *pH* of a solution of 0.104 M boric acid and 0.0217 M sodium dihydrogen borate? K_a = 5.81 x 10^{-10}

4. What is the *pH* of a 0.0273 M solution of ammonium chloride? What is the *pH* if there is 0.00572 M aqueous ammonia also present in the solution? $K_b(NH_3)$ = 1.75 x 10^{-5}

5. Use the Newton-Raphson method to solve the previous problem. $K_a(NH_4^+)$ = 5.56 x 10^{-10}

6. The Newton-Raphson method will solve weak base buffer problems if you use K_a instead of K_b. Find the pH of a solution of 0.100 M aqueous ammonia and 0.0100 M ammonium chloride by this method. Check this result by using a calculator or by using another program. $K_b(NH_3) = 1.75 \times 10^{-5}$

7. For acetic acid, test the three methods of calculating $[H^+]$ when the initial concentrations are: 4×10^{-7} M, 1×10^{-7} M, and 1×10^{-8} M. Compare these results with problem 8. $K_a = 1.79 \times 10^{-5}$

8. For phenol, test the three methods of calculating $[H^+]$ when the initial concentrations are: 7×10^{-4} M and 1×10^{-4} M. Compare these results with problem 7. $K_a = 1.05 \times 10^{-10}$

9. After a review of acid-base equilibria, what general principles would you propose with regard to the reliability of the three methods under various conditions? As an aid in answering this question, you may wish to calculate the pH of a 0.0085 M solution of phosphorous acid. $K_{a1} = 1.6 \times 10^{-2}$

17

Titration Curves

Titration curves for acid-base titrations are plots of the pH of a solution being titrated with respect to the amount of titrant that has been added. They are useful experimentally to help find indistinct endpoints, to determine the dissociation constants for weak acids and bases, and to identify different materials. Their calculation and plotting help understand theoretically what is happening during a titration, including how an endpoint can be detected with an indicator and how to select the proper indicator. (BLB Chaps. 16 and 17)

There are four distinct cases encountered during the course of a titration. They all involve an acid-base equilibrium but also depend on how much material has been reacted with respect to the amount present at the start. These four cases therefore are related to the stoichiometry of the reaction.

1. No reaction yet (before the titration starts)
2. Reaction partly complete (between start and endpoint and involves a limiting reagent problem)
3. Reaction completely finished (endpoint with the amounts in the same mole ratio as in the balanced chemical equation)
4. Excess added (past the endpoint and involves another limiting reagent problem)

Let's consider the specific case of titrating a weak acid with a strong base. The strong base is used instead of a weak base because it will result in a sharper endpoint. You may be able to see the reason for this after doing the first two sets of exercises in this chapter. Assume the stoichiometry is one to one, i.e., the acid molecule contains one H^+ to react with each molecule of base. For a weak acid, HA, being titrated with NaOH, the reaction will be:

$$HA + NaOH \rightarrow NaA + H_2O$$

This system can be described at any point during the titration using the following equilibrium expressions and equations:

1. The acid ionization equilibrium for the reaction $HA \rightarrow H^+ + A^-$

$$K_a = \frac{[H^+][A^-]}{[HA]}$$

2. The water ionization equilibrium for the reaction $H_2O \rightarrow H^+ + OH^-$

$$K_w = [H^+][OH^-]$$

3. Mass balance (C_a is the total acid concentration)

$$C_a = [A^-] + [HA]$$

4. Charge balance (sum of positive and negative ions must be equal)

$$[Na^+] + [H^+] = [OH^-] + [A^-]$$

Substitute $[HA] = [H^+][A^-] / K_a$ from equation 1 into equation 3. Then replace the two $[A^-]$ terms in the resulting equation by $[Na^+] + [H^+] - [OH^-]$ from equation 4. Then replace $[OH^-]$ in this equation with $K_w / [H^+]$ from equation 2. Multiply through by $K_a[H^+]$ to remove terms in denominators. Finally, regroup terms with like powers of $[H^+]$ to obtain the following cubic equation:

$$0 = [H^+]^3 + ([Na]+K_a)[H^+]^2 + (K_a[Na]-K_aC_a-K_w)[H^+] - K_aK_w$$

We will solve this equation using the Newton-Raphson method. We need the derivative (a calculus term) of the equation:

$$3[H^+]^2 + 2([Na]+K_a)[H^+] + (K_a[Na]-K_aC_a-K_w)$$

A new value for the [H+] is found from a starting guess using the following:

$$[H^+]_{new} = [H^+]_{old} - \frac{f([H^+]_{old})}{f'([H^+]_{old})}$$

where f([H+]) is the equation and f '([H+]) is its derivative. $[H^+]_{new}$ is put back in the equation for $[H^+]_{old}$ and another new value calculated. This process is repeated until the two [H+] values are very close to each other. The pH is then calculated.

	A	B	C	D	E
1	**Titration of a Strong or Weak Acid or Base**				
2	acid or base -->	acid			
3	Sample vol ->	50	mL	Titrant Volume ------>	
4	Sample conc ->	0.1	M	Titrant conc ------>	
5	Ka or Kb ---->	1E-05		Titrant aliquot size ->	
6					
7		pH	Titrant conc	Sample conc	f([H+])
8	0	3.0022	0	0.1	3.46E-25

	F	G	H	I
1				
2				
3	65	mL	Kw =	1E-14
4	0.1	M		
5	2	mL		
6				
7	f'([H+])	[H+]		
8	1.99E-06	0.000995		

Worksheet 17.1 Start of Sheet for Calculating a Titration Curve

Set up a sheet as shown in Worksheet 17.1. Cell B2 indicates if we are titrating and acid or base (this sheet will let us titrate either an acid or a base). Cell B3 gives the volume in mL of the acid. Cell B4 is the concentration of the acid. Cell B5 is the equilibrium constant, K_a, for the acid. Cell F3 gives the total amount of base to be added during the titration. Cell F4 gives the concentration of this base. Cell F5 gives the amount of base to be added for each point on the titration curve. Cell I3 is the value for K_w, the equilibrium constant for water ionizing.

Select Formula, Define Name ... and name the following cells:

```
ca    =$B$4
cb    =$F$4
ka    =$B$5
kw    =$I$3
va    =$B$3
```

Cell A8 is the first volume of base added. This is zero corresponding to the start of the titration. Enter the following in cell B8. This will calculate pH depending on whether an acid or base is being titrated by checking to see what is entered in cell B2. $G8 is where the $[H^+]$ will be calculated. Cancel any error messages you get because cell G8 hasn't been defined yet.

```
=IF($B$2="base",14+LOG($G8),-LOG($G8))
```

Cell C8 is the concentration of the base added after being diluted by the acid volume and equals the concentration of the base times the volume of the base added divided by the total volume of acid and base added:

```
=cb*$A8/(va+$A8)
```

Cell D8 is the concentration of the acid corrected for dilution from adding the base and equals the concentration of the acid times the volume of the acid divided by the total volume of acid and base added:

```
=ca*va/(va+$A8)
```

Cell E8 is the equation we are going to solve:

```
=$G8^3+($C8+ka)*$G8^2+(ka*$C8-ka*$D8-kw)*$G8-ka*kw
```

Cell F8 is the derivative of this equation:

```
=3*$G8^2+2*($C8+ka)*$G8+(ka*$C8-ka*$D8-kw)
```

Cell G8 is the Newton-Raphson formula for calculating the new value of the hydrogen ion from the old. The old value is the one currently in this cell and it is replaced by the new value.

```
=IF($G8<0,$B$4,$G8-$E8/$F8)
```

This cell also checks to see if the new value is negative (which is not realistic) and changes it back to a realistic value if needed.

To make the sheet do an iterative calculation of this sort, select Options, Calculation..., Iteration, OK. The calculations will keep iterating until the old and new value in cell G8 get very close together and equal the solution to the equation.

	A	B	C	D	E
1	**Titration of a Strong or Weak Acid or Base**				
2	acid or base -->	acid			
3	Sample vol ->	50	mL	Titrant Volume ------>	
4	Sample conc ->	0.1	M	Titrant conc ------>	
5	Ka or Kb --->	1E-05		Titrant aliquot size ->	
6					
7		pH	Titrant conc	Sample conc	f([H+])
8	0	3.0022	0	0.1	3.46E-25
9	2	3.6457	0.003846154	0.09615385	9.75E-27
10	4	3.9465	0.007407407	0.09259259	9.75E-27
11	6	4.138	0.010714286	0.08928571	3.29E-27
12	8	4.2818	0.013793103	0.0862069	3.29E-27
13	10	4.3992	0.016666667	0.08333333	5.77E-29
14	12	4.5003	0.019354839	0.08064516	5.77E-29
15	14	4.5905	0.021875	0.078125	5.77E-29
16	16	4.6732	0.024242424	0.07575758	5.77E-29

	F	G	H	I
1				
2				
3	65	mL	Kw =	1E-14
4	0.1	M		
5	2	mL		
6				
7	f'([H+])	[H+]		
8	1.99E-06	0.000995		
9	9.74E-07	0.000226		
10	8.65E-07	0.000113		
11	7.91E-07	7.28E-05		
12	7.27E-07	5.23E-05		
13	6.68E-07	3.99E-05		
14	6.14E-07	3.16E-05		
15	5.63E-07	2.57E-05		
16	5.16E-07	2.12E-05		

Worksheet 17.2 Part of Sheet for Calculating a Titration Curve

If you assume the following reaction with an acid whose $K_a = 1x10^{-5}$, you should get the values shown in Worksheet 17.1 using the concentrations and volumes shown.

$$HA + NaOH \rightarrow NaA + H_2O$$

To generate the rest of the points for a titration curve enter =A8+$F$5 into cell A9. Copy this cell into cells A10 through A38. Next copy cells B8 through G8 into cells B9 through G38. Then select Options, Calculate Now. This will calculate all the points for a whole titration curve. *Be sure you select this Calculate Now option each time* you change variables in the worksheet to ensure the whole titration curve is recalculated in the remainder of this chapter. Worksheet 17.2 shows the first part of this sheet. It may take a few moments to calculate all of these points, so be patient.

To make a graph of the curve, select the cells A7 through B38 and then select File, New, Chart, OK. Then select Gallery, Line, 1, OK to get a plot of all the points. Your plot should look like Chart 17.2.

Select Window, Arrange All to see both the chart and the sheet at the same time. You may need to close any other windows you have to see these best. Now change the acid concentration from 0.1 to 0.05 and observe the change on the graph. Try changing what is being titrated from an acid to a base. The equilibrium constant is now for a weak base. Notice how the curve changes. The equations for titrating a weak base are identical to titrating a weak acid except the equation solved gives OH^- concentration.

The number of points for each titration curve is fixed so care must be taken in choosing the concentration and volume of the sample and titrant so the end point of the titration will show up on the plot. Suitable values for these variables are suggested for each problem below.

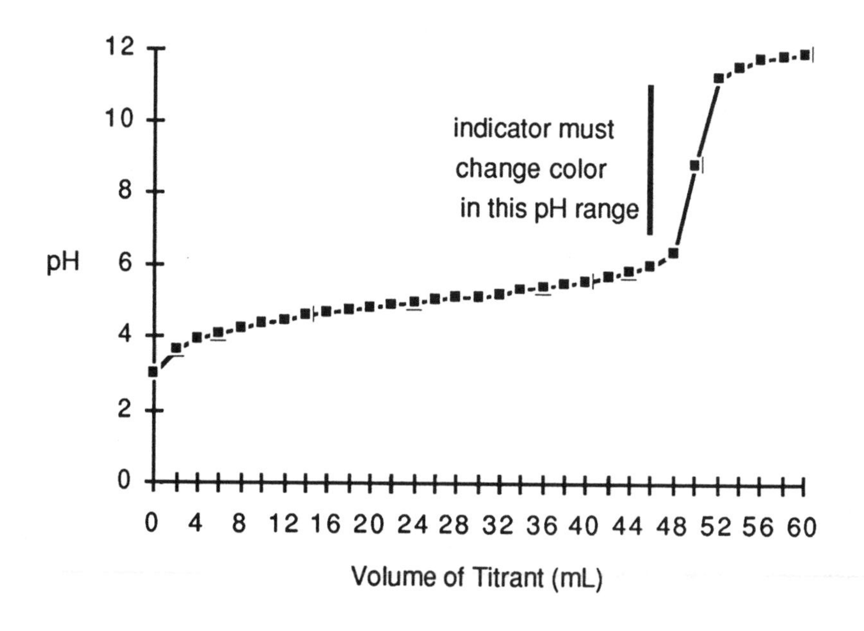

Chart 17.2 Titration Curve for Titration of a Weak Acid

If you are using an indicator to detect the endpoint of this titration, it must change color somewhere in the range indicated in Chart 17.2. Refer to Table 17.1 for examples of several indicators and over what pH range they undergo their color change. You can see that phenolphthalein or bromthymol blue would work for the titration in Chart 17.2, but methyl orange would not as it would start to change color long before the endpoint is approached.

Table 17.1 Color Change Interval of Selected Acid-Base Indicators

Indicator	pH range of color change
Methyl orange	3.1 to 4.4
Methyl red	4.2 to 6.2
Bromthymol blue	6.0 to 7.6
Phenolphthalein	8.0 to 9.8

1. How does the titration curve change with different K_a values?

Titrate 50 mL of 0.1 M acid with the following K_a values using 0.1 M base and an aliqout size of 2 mL:

a) $K_a = 1x10^{-3}$
b) $K_a = 1x10^{-6}$
c) $K_a = 1x10^{-9}$

Does the endpoint change position? Might you need a different indicator for each case? See Table 17.1.

2. How does changing the concentration and/or volume of the acid affect the curve?

Run an acid with $K_a = 1x10^{-5}$ using the following volumes and molarities:

a) 50 mL at 0.1 M
b) 25 mL at 0.1 M
c) 25 mL at 0.2 M
d) 50 mL at 0.05 M

Does the endpoint change position? What causes this to happen?

3. Does pH = 7 at the endpoint in any of the weak acid curves you have run? Why is this so?

Weak Base Modification

Change the material being titrated to a base. The worksheet will now plot the titration curves for weak bases being titrated with a strong acid such as HCl.

$$B + HCl \rightarrow HB^{+} + H_2O$$

The curves will be the mirror image of the curves for a weak acid with the same dissociation constant. Work the problems below using this worksheet.

4. How does the titration curve change with different K_b values? How is this change different than for a weak acid?

   Run 50 mL of 0.1 M base with the following K_b values:

   a) $K_b = 1x10^{-3}$
   b) $K_b = 1x10^{-6}$
   c) $K_b = 1x10^{-9}$

   Does the endpoint change position? Might you need a different indicator for each case? See Table 17.1.

5. How does changing the concentration and/or volume of the base affect the curve?

   Run a base with $K_b = 1x10^{-5}$ using the following volumes and molarities:

   a) 50 mL at 0.1 M
   b) 25 mL at 0.1 M
   c) 25 mL at 0.2 M
   d) 50 mL at 0.05 M

   Does the endpoint change position? What causes this to happen?

6. Does the pH = 7 at the end point in any of the weak base curves you have run? Why is this so?

Strong Acid Modification

To titrate a strong acid, change the material being titrated to an acid and put in an appropriately large value for K_a such as 1e+5, i.e., 100000. The reaction follows the same stoichiometry as for a weak acid.

7. Titrate 50 mL of 0.1 M strong acid. How do the indicator selection possibilities here compare with those of the weak acids? See Table 17.1.

8. How does changing the concentration and/or volume of the acid affect the curves?

Run an acid with the following volumes and molarities:

a) 50 mL at 0.1 M (same as sample curve above)
b) 25 mL at 0.1 M
c) 25 mL at 0.2 M
d) 50 mL at 0.05 M

Does the endpoint change position? What causes this to happen?

9. Does the pH = 7 at the endpoint of any of these curves? Why is this so?

Strong Base Modification

Change the material being titrated to a base. Leave the equilibrium constant value at 100000 and the calculation will be done for titrating a strong base.

10. Titrate 50 mL of 0.1 M of strong base. How do the indicator selection possibilities here compare with those of the weak bases and the strong acids? See Table 17.1.

11. How does changing the concentration and/or volume of the base affect the curves?

Titrate a strong base with the following volumes and molarities:

a) 50 mL at 0.1 M (same as curve in problem 1)
b) 25 mL at 0.1M
c) 25 mL at 0.2M
d) 50 mL at 0.05M

12. Does the pH = 7 at the endpoint of any of these curves? Why is this so?

18

Solubility Products

Solubility products are equilibrium constants for reactions involving very slightly soluble ionic materials dissolving in water. This chapter will describe some of the terminology and problems unique to this type of equilibrium. (BLB Chap. 17)

Consider an ionic compound such as AgCl dissolving in water with no other source of Ag^+ or Cl^- ions.

$$AgCl(s) \rightarrow Ag^+(aq) + Cl^-(aq)$$

Ignoring the pure solid, AgCl(s), as usual, the equilibrium expression for this reaction is :

$$K_{sp} = [Ag^+][Cl^-] = 1.8 \times 10^{-10}$$

The solubility of AgCl (i.e., the number of moles of AgCl that can dissolve in one liter) can be determined from K_{sp}, the solubility product. We can solve for the Ag^+ ion concentration by noting that since all of the Ag^+ and all of the Cl^- come from AgCl dissolving, their amounts are related to each other by the stoichiometry of the reaction. Thus, we know that $[Ag^+] = [Cl^-]$ and we can substitute this into the solubility product expression to eliminate one of the unknowns.

$$K_{sp} = [Ag^+][Cl^-] = [Ag^+][Ag^+] = [Ag^+]^2$$
$$[Ag^+] = K_{sp}^{1/2} = (1.8 \times 10^{-10})^{1/2} = 1.34 \times 10^{-5}$$

Since one Ag^+ ion is formed for each formula unit of AgCl dissolving, the solubility of AgCl will equal the $[Ag^+]$ or 0.0000134 mol/L.

The problem is a little more complex if the ratio of the two types of ions is not 1:1. Consider the case of $Cu_3(PO_4)_2$ where 3 positive ions and 2 negative ions form from dissolving each formula unit. The solubility product expression and concentration relation between the two ions is:

$$K_{sp} = [Cu^{++}]^3[PO_4^{3-}]^2 = 1.3 \times 10^{-37}$$
$$[Cu^{++}] = 3/2[PO_4^{3-}]$$

Substituting the relationship between the ion concentrations into the solubility product expression gives:

$$K_{sp} = (3/2[PO_4^{3-}])^3[PO_4^{3-}]^2 = (3/2)^3[PO_4^{3-}]^5$$
$$[PO_4^{3-}] = (K_{sp}/(3/2)^3)^{1/5} = (1.3 \times 10^{-37}/(27/8))^{1/5} = 3.29 \times 10^{-8}$$

Since two PO_4^{3-} ions are formed for each formula unit dissolving, the solubility of the compound will be one half of the phosphate ion concentration.

$$\text{solubility} = 1/2[PO_4^{3-}] = 1.64 \times 10^{-8} \text{ mol/L}$$

Set up a sheet as shown in columns A and B in Worksheet 18.1. K_{sp} will be put in cell B3. The number of positive ions formed is entered in cell B4 and the number of negative ions in B5. The solubility is calculated in cell B6 using the formula:

```
=(B3*(B5/B4)^-B5)^(1/(B4+B5))/B4
```

	A	B	C	D	E
1	Solubility product				
2					
3	Ksp--->	1.3E-37		Solubility->	1.6448E-08
4	+ ions->	3		+ ions->	3
5	- ions->	2		- ions->	2
6	Solubility->	1.64477E-08		Ksp--->	1.30014E-37

Worksheet 18.1 Sheet to Calculate Solubility and Solubility Product

Test your sheet by working the previous two examples. When it is working correctly, solve the following problems:

1. What is the solubility of the following compounds in pure water? The number following each compound is its K_{sp} value.

a) $PbSO_4$ 1.6×10^{-8}
b) $AuCl_3$ 3.2×10^{-25}

c) Cu_2S 2.5×10^{-48}
d) $Ba_3(PO_4)_2$ 3.4×10^{-23}

The K_{sp} value can be determined from the solubility by reversing the calculation. For example, if the solubility of $Cu_3(PO_4)_2$ is 1.64×10^{-8} mol/L, then since $[Cu^{++}] = 3/2[PO_4^{3-}]$, and $[PO_4^{3-}] = 1/2$ of the solubility we can get the solubility product.

$$K_{sp} = [Cu^{++}]^3[PO_4^{3-}]^2 = (3/2[PO_4^{3-}])^3[PO_4^{3-}]^2 = 27/8[PO_4^{3-}]^5$$
$$= (27/8) \times (2 \times 1.64 \times 10^{-8})^5 = 1.3 \times 10^{-37}$$

Add columns D and F to your sheet as shown in Worksheet 18.1 to work this type of problem. The formula in cell E6 is:

```
=(E3*E4)^(E4+E5)*(E5/E4)^E5
```

Check to see if your sheet is working correctly by running the previous example. Then work the following problems:

2. What is the value for K_{sp} for each of the following compounds? The solubility in mol/L is given for each case.

a) Ag_3AsO_4 1.0×10^{-22}
b) $AgBr$ 5.0×10^{-13}
c) CaF_2 3.9×10^{-11}
d) $Ca_3(PO_4)_2$ 2.0×10^{-29}

Common Ion Effect

The situation changes if the solution in which you are trying to dissolve the material already contains a significant amount of one of the ions that will form when the compound ionizes. Let's look at trying to dissolve $Zn(OH)_2$ in a solution already containing 0.100 M NaOH. The amount of OH^- ion coming from the $Zn(OH)_2$ can be ignored. *This is true only if K_{sp} is small enough and the concentration of the common ion is large enough. See Appendix G for handling this type of problem when these criteria are not met.* The $[OH^-]$ concentration is thus set equal to the NaOH concentration. Then solve for Zn^{++} ion, which will equal the solubility.

$$K_{sp} = [Zn^{++}][OH^-]^2 = [Zn^{++}](0.100)^2 = 1.2 \times 10^{-17}$$

$$[Zn^{++}] = \frac{1.2 \times 10^{-17}}{0.100^2} = 1.2 \times 10^{-15} \text{ mol/L}$$

Set up a sheet as shown in Worksheet 18.2. Enter the type of common ion in cell D3 (*pos* or *neg*). Enter the number of positive ions in D4, the number of negative ions in D5, the concentration of the common ion in D6, and the solubility product in D7. The formula in D8 calculates the solubility of the material:

```
=IF(D3="pos",(D7/D6^D4)^(1/D5)/D5,(D7/D6^D5)^(1/D4)/D4)
```

The IF checks what the charge is on the common ion and chooses the calculation accordingly. Check your sheet with the previous example. Use this sheet to work the following problems:

	A	B	C	D
1	Common Ion Problems			
2				
3	Common ion (pos or neg)-->			neg
4	Number of positive ions->			1
5	Number of negative ions->			2
6	Molarity of common ion-->			0.1
7	Solubility Product (Ksp)->			1.2E-17
8	Solubility-------------->			1.2E-15

Worksheet 18.2 Sheet for Calculating Solubility in the Presence of a Common Ion

3. What is the solubility of $PbCl_2$ in .234 M NaCl?
   $K_{sp} = 1.6 \times 10^{-5}$ for $PbCl_2$.

4. What is the solubility of $BaSO_4$ in .0581 M Na_2SO_4?
   $K_{sp} = 1.1 \times 10^{-10}$ for $BaSO_4$.

5. What is the solubility of CuS in .00837 M $Cu(NO_3)_2$?
   $K_{sp} = 6.3 \times 10^{-36}$ for CuS.

6. What is the solubility of $Ca_3(PO_4)_2$ in .0288 M $Na_3(PO_4)$?
   $K_{sp} = 2.0 \times 10^{-29}$ for $Ca_3(PO_4)_2$.

19

Entropy and Free Energy

Entropy is a measure of the disorder of a system. This property of a system depends only on its present state, which is specified by the temperature and pressure. The change in entropy that coincides with a chemical reaction may be calculated by using tables of the standard (absolute) entropies. The entropy change is positive if the overall change produces a system with more disorder. Calculation of the entropy change is analogous to the calculation of the enthalpy change. A spreadsheet for calculating the change in the enthalpy is presented in Chapter 7, Calorimeter Measurements and Standard Enthalpy of Reaction. A slightly different technique will be used with the spreadsheets developed in this chapter for computing the entropy and free energy changes that occur during a chemical reaction. The use of formula arrays is introduced for this purpose but the previous method is general and just as valid. The Gibbs function or free energy is a state function. It is of interest because the value of the free energy change predicts whether the system is at equilibrium, and if it is not at equilibrium, the direction in which the reaction will proceed. The free energy change for a reversible process is the maximum energy available, or free, to do work. The free energy change is negative if the reaction is spontaneous. The standard free energy change for a chemical reaction may be calculated from the standard free energies of the reactants and products. (BLB Chap. 19)

Entropy

Tables of the standard or absolute entropies of many chemical species exist in textbooks and reference books. Calculation of the change in the entropy of a chemical reaction is accomplished by subtracting the sum of the standard entropies of the reactants from the sum of the standard entropies of the products. The magnitude and sign of this value depends upon the balanced equation, as the standard entropies of both products and reactants are calculated as the sum of the mathematical products of the stoichiometric coefficients of each species times the standard entropy of that species. The sign of the entropy of the process or reaction is dependent upon the choice of reactants and products. This can be expressed mathematically as

$$\Delta S^\circ = \Sigma\ p \cdot S^\circ(\text{products}) - \Sigma\ r \cdot S^\circ(\text{reactants})$$

The entropy change that accompanies a chemical reaction is positive if the overall change produces a system with more disorder. If the change produces a system with more order, the value of ΔS is negative.

Worksheet 19.1 is designed to compute the entropy change for reactions that involve four or less product species and four or less reactant species. The oxidation of hydrazine with hydrogen peroxide is shown as an example of the manner in which the data are entered. For this display, the third and fourth columns for both the reactants and products have been reduced, as they are not needed in this example. The normal spreadsheet would have all of the columns for the reactants and products expanded in order to accommodate the names of the chemical species.

	A	B	C	D	E	F	G	H
1				Standard Entropy of Reaction				
2								
3		reac 1	reac 2	reac 3		prod 1	prod 2	prod 3
4	formula --->	N2H4 (l)	H2O2 (l)		------>	N2	H2O (g)	
5	coefficient -->	1	2			1	4	
6	S° (J/K-mol) =	121.2	109.6			191.5	188.7	
7								
8	Disorder			ΔS =	605.9	J/K-mol		

Worksheet 19.1 Entropy of Reaction

Formula List 19.1 shows the two cells that contain formulas. Cell E8 contains two expressions that are examples of array formulas. Array formulas are specific to Excel, and other spreadsheet software may or may not evaluate expressions of this form. To enter an array, for IBM® compatibles, you must hold down the Control and Shift keys while you press the Enter key or click on the enter box after placing the formula in cell E8. To enter an array on a Macintosh®, you must hold down the Command key when you press the Enter key or click the enter box.

IBM®, CONTROL+SHIFT+ENTER, Macintosh®, COMMAND+ENTER.

Array syntax is similar to the mathematical expression for computing the thermodynamic parameters. Computations that involve enthalpy, entropy, and free energy always use the convention of taking the sum of the products minus the sum of the reactants. Cell A8 contains a conditional statement that indicates if the system is moving toward order or disorder.

Formula List 19.1 Formulas for Worksheet 19.1

Cell A8	=IF(E8=0,"No Change",IF(E8>0,"Disorder","Order"))
Cell E8	=SUM(F5:H5*F6:H6)-SUM(B5:D5*B6:D6)

Free Energy

It is useful to define a thermodynamic quantity in terms of H and S that indicates when a process or chemical reaction is spontaneous. A chemical reaction is either spontaneous in the direction as written or nonspontaneous, which signifies that the reaction will be spontaneous in the reverse direction. The defining equation for the free energy, G, is

$$G = H - T \cdot S$$

As a reaction progresses at a specified temperature and pressure, the free energy changes as a result of the corresponding changes in the enthalpy and entropy. The resulting change in the free energy, ΔG, is

$$\Delta G = \Delta H - T \cdot \Delta S$$

This expression is valid for a reaction in general as shown above, and it is also valid when standard states are chosen. For our purposes, these standard states are represented by pure solids and liquids at one atmosphere of pressure, gases at one atmosphere of partial pressure, and solutions at 1 M concentration. The temperature may be any specified temperature, but most often it is chosen to be 298 K (25°K).

For purposes of tabulating thermodynamic data, the standard free energy of formation is used. This value is defined in a manner that is analogous to the enthalpy of formation. All of these thermodynamic values are extensive variables, and thus the tabulated values are for one mole of material. Calculations in the standard free energy change from the standard free energies of formation are completely analogous to the previous calculations for the change in the enthalpy.

$$\Delta G^\circ = \Sigma\ p \cdot \Delta G_f^\circ(\text{products}) - \Sigma\ r \cdot \Delta G_f^\circ(\text{reactants})$$

The free energy change is negative if the reaction or process is spontaneous. If the reaction is nonspontaneous, the sign of ΔG is positive. If $\Delta G = 0$, the system is at equilibrium. The implications of these statements will be expanded upon in the next chapter. For now, we will concentrate on the computation of the standard free energy change. The free energy change is the maximum energy available for conversion to work.

Worksheet 19.2 computes the free energy change for reactions that involve four or less product species and four or less reactant species. The oxidation of hydrazine with hydrogen peroxide is again shown as an example of the form in which the data are entered. For this display, the third and fourth columns for both the reactants and products have been reduced, as they are not needed in this example. The conditional statement in cell A8 and the array formula used in cell E8 are very similar to the same formulas in Worksheet 19.1.

	A	B	C	D	E	F	G	H
1			Standard Free Energy of Reaction					
2								
3		reac 1	reac 2	reac 3		prod 1	prod 2	prod 3
4	formula --->	N2H4 (l)	H2O2 (l)		------>	N2	H2O (g)	
5	coefficient -->	1	2			1	4	
6	ΔG° (kJ/mol) =	149.24	-120.4			0	-228.61	
7								
8	Spontaneous			ΔG =	-822.88	kJ/mol		

Worksheet 19.2 Free Energy of Reaction

Formula List 19.2 Formulas for Worksheet 19.2

Cell A8	=IF(E8=0,"Equilibrium",IF(E8>0,"Non-spontaneous","Spontaneous"))
Cell E8	=SUM(F5:H5*F6:H6)-SUM(B5:D5*B6:D6)

Standard Thermodynamic Quantities

Worksheet 19.3 presents the computation of all of the thermodynamic quantities that we have studied. The spreadsheet is designed to compute the entropy, enthalpy, and free energy changes for reactions that involve four or less product species and four or less reactant species. The oxidation of hydrazine with hydrogen peroxide is shown as an illustration of the form in which the data are entered. Notice that the free energy term includes a contribution from the entropy term and the enthalpy term. In this case, both of these terms work to make the reaction spontaneous. In making calculations with the thermodynamic quantities, it is imperative that you carefully use the value that corresponds to the correct physical state of the species. For this display, the third and fourth columns for both the reactants and products have been reduced, as they are not needed in this example.

Formula List 19.3 shows the cells that contain formulas from Worksheet 19.3. Cells E10, E11, and E12 contain array formulas. Again, these formulas are specific to Excel, and other spreadsheet software may or may not evaluate expressions of this form. In order to enter an array, it is necessary for you to hold down the Command key when you press the Enter key or click the enter box after you have placed the formula in the appropriate cell. Cells A10, A11, and A12 contain the conditional statements that indicate whether the system is moving toward order or disorder, endothermic or exothermic, and if the reaction is spontaneous, nonspontaneous, or at equilibrium.

	A	B	C	D	E	F	G	H
1			Standard Thermodynamic Quantities of Reaction @ 25 C					
2								
3		reac 1	reac 2	reac 3		prod 1	prod 2	prod 3
4	formula --->	N2H4 (l)	H2O2 (l)		------>	N2 (g)	H2O (g)	
5	coefficient -->	1	2			1	4	
6	S° (J/K-mol) =	121.21	109.6			191.5	188.7	
7	ΔH° (kJ/mol) =	50.63	-187.8			0	-241.8	
8	ΔG° (kJ/mol) =	149.24	-120.4			0	-228.61	
9								
10	Disorder			ΔS =	**605.89**	J/K-mol		
11	Exothermic			ΔH =	**-642.23**	kJ/mol		
12	Spontaneous			ΔG =	**-822.88**	kJ/mol		

Worksheet 19.3 Thermodynamic Quantities of Reaction

Table 19.3 Formulas for Worksheet 19.3

Cell A10	=IF(E10=0,"No Change",IF(E10>0,"Disorder","Order"))
Cell A11	=IF(E11=0,"No Change",IF(E11>0,"Endothermic","Exothermic"))
Cell A12	=IF(E12=0,"At Equilibrium",IF(E12>0,"Nonspontaneous","Spontaneous"))
Cell E10	=SUM(F5:H5*F6:H6)-SUM(B5:D5*B6:D6)
Cell E11	=SUM(F5:H5*F7:H7)-SUM(B5:D5*B7:D7)
Cell E12	=SUM(F5:H5*F8:H8)-SUM(B5:D5*B8:D8)

Problems

1. Calculate $\Delta S°$ and $\Delta G°$ for the following reactions. Predict the sign of $\Delta H°$. The standard-state values are on the next page.

$$2\ H_2\ (g) + O_2\ (g) = 2\ H_2O\ (g)$$
$$H_2\ (g) + O_2\ (g) = H_2O_2\ (g)$$
$$N_2\ (g) + O_2\ (g) = 2\ NO\ (g)$$
$$N_2\ (g) + 2\ O_2\ (g) = 2\ NO_2\ (g)$$
$$N_2\ (g) + 2\ H_2\ (g) = N_2H_4\ (g)$$

Substance	S° (J/mol-K)	ΔH°$_f$ (kJ/mol)	ΔG°$_f$ (kJ/mol)
H_2 (*g*)	130.58	0	0
O_2 (*g*)	205.0	0	0
N_2 (*g*)	191.50	0	0
NO (*g*)	210.62	90.37	86.71
NO_2 (*g*)	240.45	33.84	51.84
H_2O (*g*)	188.7	-241.8	-228.61
H_2O_2 (*g*)	232.9	-136.10	-105.48
N_2H_4 (*g*)	238.5	95.40	159.4

2. Calculate $\Delta S°$, $\Delta H°$, and $\Delta G°$ for the Haber reaction. Try to predict the signs of these three thermodynamic quantities before completing the calculations. Are your predictions verified by the spreadsheet?

$$N_2\ (g) + 3\ H_2\ (g) = 2\ NH_3\ (g)$$

Substance	S° (J/mol-K)	ΔH°$_f$ (kJ/mol)	ΔG°$_f$ (kJ/mol)
N_2 (*g*)	191.50	0	0
H_2 (*g*)	130.58	0	0
NH_3 (*g*)	192.5	-46.19	-16.66

3. Calculate $\Delta S°$, $\Delta H°$, and $\Delta G°$ for the combustion of ethyl alcohol as a liquid and as a gas. The difference between the two computed values of $\Delta H°$ for the combustion of ethyl alcohol is what important thermodynamic quantity? Which produces the largest amount of heat?

$$C_2H_5OH\ (l) + 3\ O_2\ (g) = 2\ CO_2\ (g) + 3\ H_2O\ (g)$$

Substance	S° (J/mol-K)	ΔH°$_f$ (kJ/mol)	ΔG°$_f$ (kJ/mol)
C_2H_5OH (*g*)	282.6	-235.1	-168.6
C_2H_5OH (*L*)	160.7	-277.7	-174.76
O_2 (*g*)	205.0	0	0
CO_2 (*g*)	197.9	-110.5	-137.3
H_2O (*g*)	188.7	-241.8	-228.61

4. Calculate $\Delta S°$, $\Delta H°$, and $\Delta G°$ for the following (next page):

$$C_6H_{12}O_6\ (s) + 6\ O_2\ (g) = 6\ CO_2\ (g) + 6\ H_2O\ (g)$$
$$CH_4\ (g) + 2\ O_2\ (g) = CO_2\ (g) + 2\ H_2O\ (g)$$
$$C_3H_8\ (g) + 5\ O_2\ (g) = 3\ CO_2\ (g) + 4\ H_2O\ (g)$$
$$C_8H_{18}\ (g) + 12.5\ O_2\ (g) = 8\ CO_2\ (g) + 9\ H_2O\ (g)$$
$$C_2H_2\ (g) + 2.5\ O_2\ (g) = 2\ CO_2\ (g) + H_2O\ (g)$$

Substance	S° (J/mol-K)	ΔH°$_f$ (kJ/mol)	ΔG°$_f$ (kJ/mol)
$C_6H_{12}O_6$ (*l*)	212.1	-1273.02	-910.4
CH_4 (*g*)	186.3	-74.8	-50.8
C_2H_8 (*g*)	269.9	-103.85	-23.47
C_8H_{18} (*g*)	463.67	-208.4	-17.3
C_2H_2 (*g*)	200.82	226.75	209.2
O_2 (*g*)	205.0	0	0
CO_2 (*g*)	197.9	-110.5	-137.3
H_2O (*g*)	188.7	-241.8	-228.61

5. Calculate $\Delta S°$, $\Delta H°$, and $\Delta G°$ for the following reactions. Calculate the temperature at which all of the following reactions are at equilibrium. Assume that $\Delta H°$ and $\Delta S°$ are independent of temperature.

$$0 = \Delta G° = \Delta H° - T \cdot \Delta S°$$

Compare the results within the alkaline earths. Compare the other carbonates and explain your observations. You should have values for the decomposition of six carbonates.

$$CaCO_3\ (s) = CaO\ (s) + CO_2\ (g)$$

Substance	S° (J/mol-K)	ΔH°$_f$ (kJ/mol)	ΔG°$_f$ (kJ/mol)
$CaCO_3$ (*s*)	92.88	-1207.1	-1128.76
$SrCO_3$ (*s*)	97.1	-1220	-1140
$BaCO_3$ (*s*)	112.1	-1216.3	-1137.6
$MgCO_3$ (*s*)	65.7	-1096	-1012
$PbCO_3$ (*s*)	131	-699.1	-625.5
$CdCO_3$ (*s*)	92.5	-750.6	-669.4
CaO (*s*)	39.75	-635.5	-604.17
SrO (*s*)	54.4	-592	-561.9
BaO (*s*)	70.42	-553.5	-525.1
MgO (*s*)	27.9	-597.98	-565.97
PbO (*s*)	68.7	-217.3	-187.9
CdO (*s*)	54.8	-258	-228
CO_2 (*g*)	197.9	-110.5	-137.3

20

Free Energy and Equilibrium

The Gibbs function and the equilibrium constant are thermodynamic functions that express the same thermodynamic quantity, the extent of reaction. The extent of reaction depends only on the present state of the system, which is completely specified by variables such as temperature and pressure, and either the initial or equilibrium amounts of the species. The Gibbs function is a state function. It is of interest because the value of the free energy change predicts whether the system is at equilibrium, and if it is not at equilibrium, the direction in which the reaction will proceed. The free energy change is the maximum energy available, or free, to do work. The free energy change is negative if the reaction is spontaneous. The same concept is applicable to the equilibrium constant and the reaction quotient. The reaction quotient is less than the equilibrium constant when the reaction is spontaneous. The computations presented in the preceding chapter all involved standard state free energies that uniquely specify the state of the species. In this chapter, we will develop methods for computing the free energy of a system when the species are present under nonstandard state conditions. Computing the free energy function for different concentrations of reactants and products clarifies the concept of chemical equilibrium. The spreadsheets that we will design are very similar to those presented in Chapter 7, Calorimeter Measurements and Standard Enthalpy of Reaction and Chapter 19, Entropy and Free Energy. The temperature dependence of the Gibbs function and the equilibrium constant is an important feature for understanding the effect of temperature on the extent of reaction. Alterations in the temperature and concentrations of the species may shift the direction, and they always change the extent of a chemical reaction or position of equilibrium. (BLB Chap. 19)

The Free Energy Function

In the previous chapter, the free energy was used as a criterion for the spontaneity of a reaction at constant temperature and pressure. A relationship that involved both the enthalpy and the entropy was presented in terms of standard state conditions.

$$\Delta G^\circ = \Delta H^\circ - T\cdot\Delta S^\circ$$

Not only can the standard free energy change be calculated from the change in the enthalpy and entropy, but it can be calculated from tabulated values for the standard free energies of formation of the reactants and products.

$$\Delta G^\circ = \Sigma\ p\cdot\Delta G^\circ_f(\text{products}) - \Sigma\ r\cdot\Delta G^\circ_f(\text{reactants})$$

The same physical chemical principles that provided the framework for deriving these equations also relate the free energy to the standard state free energy and any set of concentrations. This free energy function allows the calculation of the free energy under any set of conditions (temperature will be discussed later).

$$\Delta G = \Delta G^\circ + R\cdot T\cdot \ln Q$$

This expression is very useful for converting the free energy to the corresponding equilibrium constant or the reverse. When a chemical system is at equilibrium, the following conditions are true, $\Delta G = 0$ and $Q = K$. The result is

$$\Delta G^\circ = - R\cdot T\cdot \ln K$$

This log function is useful for converting the equilibrium constant to the free energy. The exponential form of the equation is more useful for the conversion of free energy to the corresponding equilibrium constant.

$$K = e^{-\Delta G^\circ/R\cdot T}$$

ΔG as a Function of Concentration

We will investigate the profile of ΔG as a reaction progresses to and through the conditions that represent the system at equilibrium. This is a hypothetical situation, as the reaction will always proceed to the equilibrium conditions and not beyond those conditions. We could choose to begin with only reactants present that would produce a large negative value for the free energy. As the reaction progresses, the large negative value for ΔG would decrease until the value is zero. These particular conditions would describe the system when it is at equilibrium. If we continue our calculations past this point, the value of ΔG becomes increasingly positive, which signifies that with these concentrations the reaction would proceed to produce a higher level of reactant concentrations and thereby reduce the level of the product concentrations. In essence, the profile of ΔG proceeds from $-\infty$ through zero to $+\infty$ as the reaction proceeds from having only reactants present to the condition of having equilibrium concentrations and then on to the conditions of having only products present.

In Chapter 15, Chemical Equilibrium, a spreadsheet for computing the equilibrium concentration of nitrogen, hydrogen, and ammonia for the Haber reaction was developed. This spreadsheet used a value of $K = 4.51 \times 10^{-5}$ atm^{-2} at a temperature of 450°C, which corresponds to a value of $\Delta G^\circ = 60.166$ kJ. With the initial concentrations of nitrogen, hydrogen, and ammonia at 1 atm, 3 atm, and 2 atm of partial pressure, the resulting equilibrium partial pressures are 1.935 atm, 5.804 atm, and 0.131 atm. These results are summarized:

$$N_2\,(g) \;+\; 3\,H_2\,(g) \;=\; 2\,NH_3\,(g)$$

	N_2	H_2	NH_3
Initial	1 atm	3 atm	2 atm
Equilibrium	1.935	5.804	0.131

Worksheet 20.1 is designed to compute the free energy changes for this reaction starting with the conditions of 1.9 atm, 5.7 atm, and 0.2 atm for the partial pressures of nitrogen, hydrogen, and ammonia. Succeeding values of ΔG are calculated using an increment of 0.005 atm as the increase in the partial pressure nitrogen, a value of three times the increment for the increase in the partial pressure of hydrogen, and a value of two times the increment for the decrease in the partial pressure of ammonia. This calculation represents a closed system with these three gases present. The initial conditions represent a system that is not at equilibrium, and that will produce an increase in the partial pressures of the reactants and a corresponding decrease in the partial pressure of the product as it moves toward equilibrium. The calculation proceeds through a value of ΔG that is very close to equilibrium and continues to values that represent a condition where the system would have to decrease the partial pressures of the reactants and increase the partial pressure of the product in order to achieve equilibrium.

Using this spreadsheet, compute the values of the free energy when the initial values of nitrogen, hydrogen, and ammonia are .95, 2.85, and 0.1 for partial pressures in atmospheres. This represents a system with a reduced total pressure. Does LeChatelier's Principle predict the results?

	A	B	C	D	E
1	Standard Free Energy and Equilibrium				
2					
3		N2 (g) +	H2 (g) -->	NH3 (g)	
4	initial P -->	1.9	5.7	0.2	
5	increment -->	0.005			
6		<--------------P (atm) -------------->			
7	direction	N2 (g)	H2 (g)	NH3 (g)	ΔG (kJ)
8	react<-	1.9	5.7	0.2	5.56
9	react<-	1.905	5.715	0.19	4.88
10	react<-	1.91	5.73	0.18	4.16
11	react<-	1.915	5.745	0.17	3.41
12	react<-	1.92	5.76	0.16	2.62
13	react<-	1.925	5.775	0.15	1.78
14	react<-	1.93	5.79	0.14	0.89
15	->prod	1.935	5.805	0.13	-0.06
16	->prod	1.94	5.82	0.12	-1.09
17	->prod	1.945	5.835	0.11	-2.19
18	->prod	1.95	5.85	0.1	-3.40
19	->prod	1.955	5.865	0.09	-4.73
20	->prod	1.96	5.88	0.08	-6.21

Worksheet 20.1 Free Energy of Reaction

Formula List 20.1 shows the cells that contain crucial formulas for calculating the free energy with Worksheet 20.1. With the Fill Down command under Edit on the menu bar, cells A9 through A20 are filled with contents from Cell A8. In a similar fashion, cells 10 through 20 in columns B, C, and D are filled with the respective contents of cells B9, C9, and D9. The spreadsheet is completed by filling cells E9 through E20 with contents of cell E8. The Gibbs function can be investigated by altering the initial partial pressures and increment in cells B4, C4, D4, and B5.

Formula List 20.1 Formulas for Worksheet 20.1

```
Cell B5    Defined as x
Cell A8    =IF(E8>0,"react<-","->prod")
Cell B8    =B4
Cell C8    =C4
Cell D8    =D4
Cell E8    =60.165+.0083145*723.15*LN(D8^2/(B8*C8^3))
Cell B9    =B8+x
Cell C9    =C8+3*x
Cell D9    =D8-2*x
```

Free Energy and the Equilibrium Constant

Worksheet 20.2 can be used as the basis for the design of a spreadsheet for calculating the equilibrium constant from the standard state free energies of formation of the products and reactants. For our purposes, these standard states are represented by: pure solids and liquids at one atmosphere of pressure, gases at one atmosphere of partial pressure, and solutions at 1 M concentration. The temperature may be any specified temperature, but most often it is chosen to be 298 K (25°C).

$$K = e^{-\Delta G^\circ / R \cdot T}$$

Worksheet 20.2 computes the change in free energy for reactions of four or less product species and four or less reactant species. The dissociation of water into hydrogen ion and hydroxide ion is shown as an example. For this display, the third and fourth columns for both the reactants and products have been reduced, as they are not needed in this example.

	A	B	C	D	E	F	G	H
1		Standard Free Energy & The Equilibrium Constant						
2								
3		reac 1	reac 2	reac 3		prod 1	prod 2	prod 3
4	formula --->	H2O (l)			------>	H+ (aq)	OH- (aq)	
5	coefficient -->	1				1	1	
6	ΔG° (kJ/mol) =	-237.18				0	-157.29	
7								
8				ΔG° =	79.89	kJ/mol		
9								
10				K=	1.01E-14			

Worksheet 20.2 Computing the Equilibrium Constant

Formula List 20.2 shows the cells that contain formulas. Cell E8 contains array formulas. Arrays are specific to Excel, and other spreadsheets may not evaluate expressions of this form. To enter a formula array with an IBM®, CONTROL + SHIFT + ENTER, and with a Macintosh®, COMMAND + ENTER after you have placed the formula in the cell.

Table 20.2 Formulas for Worksheet 20.2

Cell E8	=SUM(F5:H5*F6:H6)-SUM(B5:D5*B6:D6)
Cell E10	=EXP(-E8/.0083145*298.15)

The Free Energy Under Any Conditions

Worksheet 20.2 can be used as the basis for developing a spreadsheet for calculating the equilibrium constant from the standard state free energies of formation and the value of Q with the corresponding ΔG. A row for the concentration or partial pressure of all of the species has to be added. All of the concentration values must be a finite nonzero number for the proper function of this spreadsheet. The cells are initially filled with ones until you enter a specific concentration in molarity or a partial pressure in atmospheres. The temperature may be any specified temperature that is entered in cell B11. Worksheet 20.3 illustrates the computation of the free energy under any conditions of temperature and concentration. Formula List 20.3 shows the cells that contain formulas in Worksheet 20.3.

	A	B	C	D	E	F	G	H
1				Standard Free Energy & Equilibrium				
2								
3		reac 1	reac 2	reac 3		prod 1	prod 2	prod 3
4	formula --->	H2O (l)			------>	H+ (aq)	OH- (aq)	
5	coefficient -->	1				1	1	
6	ΔG° (kJ/mol) =	-237.18				0	-157.29	
7	Conc (M or atm) ->	1	1	1		1E-07	1E-07	1
8								
9				ΔG° =	**79.89**	kJ/mol		
10				K =	1.009E-14			
11	Temp (K) --->	298.15		Q =	1E-14			
12				ΔG =	-0.022492	kJ/mol		

Worksheet 20.3 Computing the Free energy

Formula List 20.3 Formulas for Worksheet 20.3

Cell E9	=SUM(F5:H5*F6:H6)-SUM(B5:D5*B6:D6)
Cell E10	=EXP(-E9/.0083145*T)
Cell E11	=PRODUCT(F7:H7^F5:H5)/PRODUCT(B7:D7^B5:D5)
Cell F13	=G0+.0083145*T*LN(Q)
Cell B11	Defined as T
Cell E9	Defined as G0
Cell E11	Defined as Q

The array in cell E9 requires holding down the Control and Shift key while depressing the Enter key on an IBM® and the Command and Enter key for the Macintosh® after you have placed the formula in the cell. The standard free energy change, ΔG°, is defined as G0, where the latter symbol is a zero not an O. Cell E11 contains another example of an array formula that multiplies a series of concentrations raised to the power that represents the stoichiometric coefficient in the balance equation. The convention for placing the products in the numerator and the reactants in the denominator when calculating Q is the same as for K.

Temperature Dependence of ΔG° and K

The temperature dependence of ΔG° can be obtained from

$$\Delta G^\circ = \Delta H^\circ - T \cdot \Delta S^\circ$$

with the assumption that ΔS° and ΔH° are independent of the temperature. This assumption is useful in some cases, although it is not valid in general. The temperature dependence of K can be obtained by combining the previous equation with

$$\Delta G^\circ = - R \cdot T \cdot \ln K$$

which yields

$$\ln K = - \frac{\Delta H^\circ}{R \cdot T} + \frac{\Delta S^\circ}{R}$$

A graph of ln K as the vertical axis and $1/T$ as the horizontal axis produces an intercept with the value, $\Delta S^\circ/R$, and a slope with a value of - $\Delta H^\circ/R$. When ΔH° has a temperature dependence, the curve is fit in a manner that produces an average value for ΔH° over the temperature range that is considered. The determination of the equilibrium constant at two different temperatures produces two equations that may then be subtracted. This results in an equation that is useful for computing the equilibrium constant at a new temperature when its value is known at a specified temperature.

$$\ln \frac{K_2}{K_1} = - \frac{\Delta H^\circ}{R} \left(\frac{1}{T_2} - \frac{1}{T_1} \right)$$

The exponential form of this equation was used to compute the value of the equilibrium constant at different temperatures in Chapter 15, Chemical Equilibrium. A related form, the Clausius-Clapeyron equation, was used in Chapter 13, Vapor Pressure.

Worksheet 20.4 computes the ΔH°, ΔG°, K, and the reaction quotient at a specified temperature. The standard free energy and the equilibrium constant are then calculated for a different temperature. These results are the mathematical equivalent of LeChatelier's principle.

	A	B	C	D	E	F	G	H
1		**Standard Thermodynamic Quantities of Reaction @ 25 C**						
2								
3		reac 1	reac 2	reac 3		prod 1	prod 2	prod 3
4	formula --->	H2O (l)			------>	H+ (aq)	OH- (aq)	
5	coefficient -->	1				1	1	
6	ΔH° (kJ/mol) =	-285.83				0	-229.99	
7	ΔG° (kJ/mol) =	-237.18				0	-157.29	
8	Conc (M or atm)->	1	1	1		1.2E-07	1.2E-07	1
9								
10	T1 (K) =	**298.15**		ΔH° =	55.84	kJ/mol		
11				ΔG° =	**79.89**	kJ/mol		
12	**reactant<-**			K =	**1.01E-14**			
13				Q =	1.44E-14			
14								
15	T2 (K) =	**313.15**		ΔG° =	**81.10**	kJ/mol		
16				K =	**2.97E-14**			
17	**->product**			Q =	1.44E-14			

Worksheet 20.4 Temperature Dependence of ΔG° and K

Worksheet 20.4F illustrates the formulas needed for this spreadsheet. Again, ΔG° is represented by G0 and the reference temperature by T0. Both of these quantities employ a zero and not the letter O.

	E
10	=SUM(F5:H5*F6:H6)-SUM(B5:D5*B6:D6)
11	**=SUM(F5:H5*F7:H7)-SUM(B5:D5*B7:D7)**
12	**=EXP(-G0/(0.0083145*T0))**
13	=PRODUCT(F8:H8^F5:H5)/PRODUCT(B8:D8^B5:D5)
14	
15	**=T*(G0-H)/T0+H**
16	**=EXP(-G/(0.0083145*T))**
17	=PRODUCT(F8:H8^F5:H5)/PRODUCT(B8:D8^B5:D5)

Worksheet 20.4F Formulas for Worksheet 20.4

Formula List 20.4 Formulas for Worksheet 20.4

Cell B10	Defined as T0
Cell B15	Defined as T
Cell E10	Defined as H
Cell E11	Defined as G0
Cell E15	Defined as G

Problems

1. Using Worksheet 20.1 as a prototype, investigate the changes in the free energy of a solution of molecular acetic acid and the ions formed by its dissociation. For the equilibrium constant of acetic acid at 25°C, use $K_a = 1.8 \times 10^{-5}$. Select any initial concentration for the molecular species and then use values for $H^+(aq)$ and acetate ion to compute values of ΔG on both sides of the equilibrium conditions. Calculations of the equilibrium concentrations were presented in Chapter 15, Chemical Equilibrium.

2. Investigate the changes in the free energy for a solution of molecular nitrous acid and the ions formed by its dissociation. For the equilibrium constant of nitrous acid at 25°C, use $K_a = 1.8 \times 10^{-5}$. Select any initial concentration for the molecular species and then use values of $H^+(aq)$ and nitrite ion to compute values of ΔG on both sides of the equilibrium conditions.

3. Investigate the changes in the free energy for a solution of molecular hydroxylamine and the ions formed by its dissociation. For the equilibrium constant of hydroxylamine at 25°C, use $K_b = 1.1 \times 10^{-8}$. Select any initial concentrations that produce chemical equilibrium. Then compute values of ΔG on either side of the chemical equilibrium.

4. Investigate the changes in the free energy function for partial pressures that produce both spontaneous and nonspontaneous reactions for the chemical equilibrium of $I_2(g)$, $Br_2(g)$, and $IBr(g)$.

$$I_2(g) + Br_2(g) = IBr(g) \quad \text{where } K_p(298\text{K}) = 7.66$$

   Select a set of initial partial pressures for the molecular species that produce an equilibrium situation and then compute values of ΔG on both sides of these equilibrium conditions.

5. Investigate the changes in the free energy function for partial pressures that produce both spontaneous and nonspontaneous reactions for the chemical equilibrium between $N_2O_4(g)$ and $NO_2(g)$.

$$N_2O_4(g) = 2\ NO_2(g) \quad \text{where } K_p(373\text{K}) = 6.49$$

   Select a set of initial partial pressures for the molecular species that produce an equilibrium situation and then compute values of ΔG on both sides of these equilibrium conditions.

6. Investigate the changes in the free energy function for partial pressures that produce both spontaneous and nonspontaneous reactions for the chemical equilibrium of $PCl_5(g)$, $PCl_3(g)$, and $Cl_2(g)$.

$$PCl_5(g) = PCl_3(g) + Cl_2(g) \quad \text{where } K_p(433K) = 0.750$$

Select a set of initial partial pressures for the molecular species that produce an equilibrium situation and then compute values of ΔG on both sides of these equilibrium conditions.

7. What is the value of the equilibrium constant at 500 K and 200 K for

$$H_2O_2(g) = H_2O(g) + 1/2\ O_2(g)$$

if at 298 K, $\Delta G° = -123.3$ kJ and $\Delta H° = -106$ kJ?

8. What is ΔH for $N_2O_4(g) = 2\ NO_2(g)$ if $K_p(298K) = 0.114$ and $K_p(373K) = 6.49$? What is $K_p(200K)$? What is $K_p(500K)$?

9. Calculate K_{sp}, for the four halide salts of silver at 25°C and 40°C. Find concentrations that produce a precipitate at one and not the other of these temperatures. The solubility is higher at which temperature?

$$AgX(s) = Ag^+(aq) + X^-(aq)$$

Substance	S° (J/mol-K)	ΔH°$_f$ (kJ/mol)	ΔG°$_f$ (kJ/mol)
$AgF(s)$	-84	-203	-185
$AgCl(s)$	96.11	-127.03	-109.72
$AgBr(s)$	107.1	-99.5	-95.939
$AgI(s)$	114	-62.38	-66.32
$Ag^+(aq)$	73.93	105.9	77.111
$F^-(aq)$	-9.6	-329.1	-276.5
$Cl^-(aq)$	55.1	-167.46	-131.17
$Br^-(aq)$	80.71	-120.9	-102.82
$I^-(aq)$	109.4	-55.94	-51.67

10. Calculate $K_p(298K)$ and $K_p(721K)$ for the formation of hydrogen iodide. Predict whether the production is favored at high or low temperature. Are your predictions verified by the spreadsheet?

$$H_2(g) + I_2(g) = 2\ HI(g)$$

Substance	S° (J/mol-K)	ΔH°$_f$ (kJ/mol)	ΔG°$_f$ (kJ/mol)
$H_2(g)$	130.58	0	0
$I_2(g)$	260.57	62.25	19.37
$HI(g)$	206.3	25.94	1.30

11. Calculate K_{sp}, for the two halide salts of lithium and calcium at 25°C and 40°C. When is the salt more soluble at the higher temperature? If possible, find concentrations that produce a precipitate at one and not the other of these temperatures.

$$LiX(s) = Li^+(aq) + X^-(aq)$$

Substance	S° (J/mol-K)	ΔH°$_f$ (kJ/mol)	ΔG°$_f$ (kJ/mol)
$LiF(s)$	-84	-203	-185
$LiCl(s)$	96.11	-127.03	-109.72
$CaF_2(s)$	107.1	-99.5	-95.939
$CaCl_2(s)$	114	-62.38	-66.32
$Li^+(aq)$	73.93	105.9	77.111
$Ca^{2+}(aq)$	73.93	105.9	77.111
$F^-(aq)$	-9.6	-329.1	-276.5
$Cl^-(aq)$	55.1	-167.46	-131.17

12. Calculate the equilibrium constant for the decomposition of the alkaline carbonates at temperatures of 300°C, 900°C, and 1300°C. Compare the results within the alkaline earths.

$$MCO_3\ (s) = MO\ (s) + CO_2\ (g)$$

Substance	S° (J/mol-K)	ΔH°$_f$ (kJ/mol)	ΔG°$_f$ (kJ/mol)
$CaCO_3$ (s)	92.88	-1207.1	-1128.76
$SrCO_3$ (s)	97.1	-1220	-1140
$BaCO_3$ (s)	112.1	-1216.3	-1137.6
$MgCO_3$ (s)	65.7	-1096	-1012
CaO (s)	39.75	-635.5	-604.17
SrO (s)	54.4	-592	-561.9
BaO (s)	70.42	-553.5	-525.1
MgO (s)	27.9	-597.98	-565.97
CO_2 (g)	197.9	-110.5	-137.3

21

Electrochemistry

The cell potential or emf, the free energy, and the equilibrium constant are thermodynamic functions that express the same quantity, the extent of reaction. The cell potential is the sum of the standard electrode potentials of the two half-reactions when the least positive reduction potential is written as an oxidation potential and the other as a reduction potential. Thus, a spontaneous process has a positive cell potential that corresponds to a negative change in the free energy. The free energy is equal to the negative of the product of the moles of electrons transferred, the Faraday, and the cell potential. Standard half-cell potentials tabulated as reduction potentials can be used to compute the voltage of a cell operating under standard state conditions. The Nernst Equation relates the standard cell potential and the concentrations of the species to the voltage produced by the cell under the particular set of nonstandard state conditions. The response of voltage to the concentration or partial pressure of the chemical species offers an analytical tool for determining the concentration or partial pressure of a selected species. The cell potential can be used in a manner that is analogous to pH for determining the end-point of redox titrations. Redox titrations are similar to the titrations of Chapter 16, Acid-Base Equilibria. A potentiometric precipitation titration illustrates the use of the cell potential for determining the end-point of a chemical reaction. (BLB Chap. 20)

emf and the Equilibrium Constant

The emf produced by a electrochemical cell is designated by the symbol E. When the cell is operated under standard state conditions, the emf produced is the standard cell potential, $E°$. Standard half-cell potentials are tabulated by convention as reduction potentials with the arbitrary choice of zero for the $H^+(1\ M) \mid H_2(1\ atm)$ half-reaction. The cell potential can be calculated from the two half-reactions by reversing the least positive half-cell reduction potential and adding it to the other half-cell reduction potential.

$$E°_{cell} = E°_{ox} + E°_{red}$$

As a result, the overall chemical reaction has a species that is oxidized and a species that is reduced. The electron(s) are transferred from the species that is oxidized to the species that is reduced. This procedure results in a positive cell potential that corresponds to a spontaneous process. This is the description of the chemical system when the cell is operating as a battery. The free energy, which must be negative when the process is spontaneous, is given by

$$w_{max} = \Delta G^\circ = -n \cdot F \cdot E^\circ$$

The free energy is given in joules when the standard cell potential is in volts and $F = 96{,}485$ coulombs/mole. The change in the free energy for the chemical reaction represents the maximum work that the system can provide when the process is carried out in a reversible manner.

Physical chemical principles provide a basis for deriving an equation that relates the standard state cell potential to the equilibrium constant at a designated temperature. This equation is a special case of a more general expression that will be presented in the next section.

$$E^\circ = \frac{R \cdot T}{n \cdot F} \cdot \ln K$$

This equation facilitates the conversion of a corresponding value for the equilibrium constant to the standard state cell potential. The exponential form of this equation is more convenient for converting the corresponding value of the standard state cell potential to the equilibrium constant.

$$K = e^{n \cdot F \cdot E^\circ / R \cdot T}$$

Worksheet 21.1 computes the standard state cell potential, E°, the standard state free energy, ΔG°, and the equilibrium constant, K, for oxidation-reduction reactions. Use of this spreadsheet requires values for the number of moles of electrons transferred in the balanced equation, the half-cell potential of the species that is oxidized and the half-cell potential of the species that is reduced. When a half-reaction is reversed to produce an oxidation potential instead of a reduction potential, the sign of the half-cell potential is reversed or changed. The thermodynamic values may be computed at any selected temperature, although the half-cell potentials change when the temperature is changed. We will ignore the temperature dependence of the electrochemical thermodynamic variables. It should be observed that very small positive values for the cell potential signify rather large equilibrium constants and correspondingly large negative values for the free energy. All of these thermodynamic parameters can be used for determining if a particular reaction is spontaneous.

	A	B	C	D
1	Cell Potential --> Free Energy & K			
2				
3	8.31452	<-R (J/K-mol)		
4	96485.3	<-F (C/mol)		
5				
6	T (K) =	298.15		
7	n (mol) =	1		
8	E° (ox) =	-0.521		Cu -> Cu+
9	E° (red) =	0.153		Cu++ -> Cu+
10	E° =	-0.368	V	
11	ΔG° =	35,507	J	
12	K =	6.02E-07		

Worksheet 21.1 E° Converted to ΔG° and K

Worksheet 21.1F illustrates the crucial formulas that are entered in the cells that produce computed results. Formula List 21.1 documents those cells that are designated by defined characters in Worksheet 21.1.

	B
10	=B8+B9
11	=-n*F*E0
12	=EXP(n*F*E0/(Rg*T))

Worksheet 21.1F Formulas for Worksheet 21.1

Formula List 20.1 Formulas for Worksheet 20.1

Cell A3	Defined as Rg
Cell A4	Defined as F
Cell B6	Defined as T
Cell B7	Defined as n
Cell B10	Defined as E0

Worksheet 21.2 illustrates the computation of the solubility product of AgBr(*s*) using the reduction potentials of silver ion and of silver bromide. The convention for the solubility product requires that the chemical reaction have the slightly soluble ionic material, AgBr(*s*), as the reactant, and $Ag^+(aq)$ and $Br^-(aq)$ as products. The reaction when written in this direction is not spontaneous; thus the cell potential is negative.

$$AgBr(s) = Ag^+(aq) + Br^-(aq)$$

	A	B	C	D
1	Cell Potential --> Free Energy & K			
2				
3	8.31452	<-R (J/K-mol)		
4	96485.3	<-F (C/mol)		
5				
6	T (K) =	298.15		
7	n (mol) =	1		
8	E° (ox) =	-0.799		Ag -> Ag+
9	E° (red) =	0.071		AgBr -> Ag + Br-
10	E° =	-0.728	V	
11	ΔG° =	70,241	J	
12	K =	**4.95E-13**		

Worksheet 21.2 Solubility Product of AgBr(s)

Dependence of emf On Concentration

The emf produced by a chemical reaction depends on the concentrations of the ions in solution and on the partial pressure of the gases. This in turn provides chemists with a very useful analytical method for determining ion concentrations and partial pressures. Measurement of the emf and the use of the appropriate form of the Nernst Equation allow for the calculation of ion concentrations or partial pressures. The basis of the Nernst Equation is

$$\Delta G = \Delta G^\circ + RT \cdot \ln Q$$

The reaction quotient, Q, has the form of the equilibrium constant, except that the concentrations and partial pressures represent any conditions, not just those at equilibrium. When the ion concentrations and partial pressures of the gases are such that the system is at equilibrium, the equation becomes

$$\Delta G^\circ = - RT \cdot \ln K$$

The more general equation for the free energy at any ion concentrations and any partial pressures may be transformed by the relationship between the free energy and the cell potential. This is the Nernst Equation, which will provide a solution to the cell potential at nonstandard state conditions.

$$E_{\text{cell}} = E^\circ_{\text{cell}} - \frac{R \cdot T}{n \cdot F} \cdot \ln Q$$

Worksheet 21.3 is designed to compute several values of interest for reactions that involve up to three product species and three reactant species. This spreadsheet could be altered to calculate the cell potential for a reaction that has more than three product and three reactant species if necessary. As before, in most cases the temperature is 298.15 K, which is precisely the temperature at which most cell and half-cell potentials are determined. The number of moles of electrons transferred which relates to the balanced chemical equation must be specified. The half-cell potential of the species that is oxidized and that of the material that is reduced must also be provided. The spreadsheet then computes several parameters that are associated with thermodynamic equilibrium and several that relate to nonstandard state conditions if these are appropriate. Worksheet 21.3F shows the cells that contain formulas in Worksheet 21.3. Cell D14 contains two array formulas that multiply a series of concentrations raised to the power that represents the stoichiometric coefficient in the balance equation. The convention for placing the products in the numerator and the reactants in the denominator when calculating Q is the same as for K. An array formula requires that the Command key be held down when you press the Enter key or click the enter box.

	A	B	C	D	E	F	G	H
1			Cell Potential for Nonstandard State Conditions					
2								
3		reac 1	reac 2	reac 3		prod 1	prod 2	prod 3
4	formula --->	Cu2+(aq)	Cu(s)		------>	Cu+(aq)		
5	coefficient -->	1	1			1		
6	Conc (M or atm)->	0.01	0.01	1		0.1	1	1
7								
8	T (K) =	298.15		8.3145	<- R (J/K-mol)			
9	n (mol) =	1		96485	<- F (C/mol)			
10	E° (ox) =	0.521						
11	E° (red) =	0.153						
12	E° =	**0.674**						
13	ΔG° =	-65,031						
14	K =	2.5E+11	Q =	1.0E+03	**Q<K**			
15	ΔG =	-47,907			**->prod**			
16	E =	**0.497**			**spontaneous**			

Worksheet 21.3 Reaction of Cu^{2+}(aq) and Cu(s)

LeChatelier's Principle can be applied to these reactions to predict whether a given change in the concentration of a species will increase or decrease the cell potential. In this case, the concentrations of the reactants were reduced more than the concentration of the product, and thus the cell potential under these conditions is less than it is when the cell is operated at standard state conditions. The cell potential, the free energy, and the equilibrium constant are measures of the same thermodynamic parameter.

	B
12	=B10+B11
13	=-n*F*E0
14	=EXP(n*F*E0/(Rg*T))
15	=-n*F*E
16	=E0-Rg*T*LN(Q)/(n*F)

	D
14	=PRODUCT(F6:H6^F5:H5)/PRODUCT(B6:D6^B5:D5)

	E
14	=IF(Q>K,"Q>K","Q<K")
15	=IF(Q>K,"react<-","->prod")
16	=IF(Q>K,"nonspontaneous","spontaneous")

Worksheet 21.3F Formulas for Worksheet 21.3

Formula List 21.3 lists the cells defined by alphanumeric characters. These shorten the entries and result in equations that more closely represent the usual notation used in chemistry. Calculations with this spreadsheet will illustrate that small values for the voltage of a cell (0.2 volts) still produce very large equilibrium constants. Changes in concentration have a very reduced effect on the cell potential, as predicted by the Nernst Equation.

Formula List 21.3 Formulas for Worksheet 21.3

Cell B8	Defined as T
Cell B9	Defined as n
Cell B12	Defined as E0
Cell B14	Defined as K
Cell B16	Defined as E
Cell D8	Defined as Rg
Cell D9	Defined as F
Cell D14	Defined as Q

The extent of reaction as determined by the equilibrium position and by the change in free energy have been discussed in Chapters 15 and 20. Another useful approach toward understanding the consequences of these thermodynamic variables is offered by investigating the changes in the cell potential as the concentrations of the reaction species are varied. The oxidation of copper(*s*) by silver(I) to form silver(*s*) and copper(II)

$$Cu(s) + 2\,Ag^{+}(aq) = Cu^{2+}(aq) + 2\,Ag(s)$$

will serve to illustrate the changes in the voltage as the concentration of the two ions is varied. The silver-copper cell has a cell potential of 0.462 volts when copper(II) and silver(I) are at unit activity. The condition of unit activity is approached when both solutions are at 1.0 molar concentration. Worksheet 21.4 emphasizes the similarity of the cell potential and the free energy by computing both as the concentrations of the reactants and products are changed in accord with the chemical stoichiometry. The particular choice of the initial and final concentrations is arbitrary, but constrained by the stoichiometry. The initial concentrations are chosen to be 1.0 M in both species of ions. The silver(I) ion concentration is reduced by the term labeled factor, which is set at ten. The form of the Nernst Equation shows that the cell voltage is not a sensitivity function of the concentration terms unless they happen to be raised to a large power.

	A	B	C	D
1	Cu(s) + 2 Ag+(aq) = Cu2+(aq) + 2 Ag(s)			
2				
3	E° (V) =	0.462		
4	initial [Cu2+] =	1		
5	initial [Ag+] =	1		
6	factor =	10	<-- reduction of [Ag+]	
7				
8	[Ag+]	[Cu2+]	E(volts)	ΔG(kJ)
9	1	1	0.4620	-89.15
10	0.1	1.45	0.3980	-76.81
11	0.01	1.495	0.3384	-65.31
12	0.001	1.4995	0.2792	-53.88
13	0.0001	1.49995	0.2200	-42.45
14	0.00001	1.499995	0.1608	-31.03
15	0.000001	1.4999995	0.1016	-19.60
16	0.0000001	1.49999995	0.0424	-8.18
17	0.00000001	1.5	-0.0168	3.24
18	0.000000001	1.5	-0.0760	14.67
19	1E-10	1.5	-0.1352	26.09
20	1E-11	1.5	-0.1944	37.52

Worksheet 21.4 The Reaction of Cu(s) and Ag^{+}(aq)

Worksheet 21.4F details the formulas and Table 21.4 lists the defined cells in this spreadsheet . Initial values of 1.4999995 M for the copper(II) and 1.0 x 10^{-6} M for silver(I) with a factor of two for decreasing the silver(I) ion concentration focus on the conditions necessary for equilibrium. Further changes in the initial concentrations of the two ions show that the concentrations needed to achieve equilibrium are 1.5 M for copper(II) and 1.92 x 10^{-8} M for silver(I).

	A	B	C	D
8	[Ag+]	[Cu2+]	E(volts)	ΔG(kJ)
9	=A0	=C0	=E0-0.0296*LOG10(B9/A9^2)	=-2*96.485*C9
10	=A9/f	=B9+(A9-A10)/2	=E0-0.0296*LOG10(B10/A10^2)	=-2*96.485*C10
11	=A10/f	=B10+(A10-A11)/2	=E0-0.0296*LOG10(B11/A11^2)	=-2*96.485*C11

Worksheet 21.4F Formulas for Worksheet 21.4

Table 21.4 Formulas for Worksheet 21.4

Cell B3	Defined as E0
Cell B4	Defined as A0
Cell B5	Defined as C0
Cell B6	Defined as f

A Potentiometric Precipitation Titration

A potentiometric titration of chloride ion with silver ion demonstrates the behavior of the cell potential when the concentration of silver(I) ion is changing. A very substantial change in the cell potential at the equivalence point serves as a useful indicator for the analytical chemist. For this titration, a reference electrode is necessary in order to have a voltage to measure, and for this purpose a calomel electrode with a standard potential of 0.268 V is used. A diagram of the experimental arrangement is presented in Figure 21.1. The known or unknown solution of chloride ion is present in the beaker and it is called the analyte. The silver(I) solution is the titrant with a known concentration and it is added to the beaker in measured volumes with the aid of a buret.

$$Cl^-(aq) + Ag^+(aq) = AgCl(s)$$

This titration has three distinct regions in terms of the computation of the silver(I) ion concentration. Before the equivalence point, the solution in the beaker contains an excess of chloride ion. The concentration of silver(I) ion is computed by using the solubility product expression with the excess chloride ion dictating the low concentration of silver(I) ion.

$$[Ag^+] = K_{sp}/[Cl^-]$$

The equivalence point is the second region and $[Ag^+]$ is determined by the solubility of the precipitate, silver chloride.

$$[Ag^+] = \sqrt{K_{sp}}$$

The third region is that portion of the titration that occurs after the equivalence point is achieved. In this region, the concentration of silver(I) increases dramatically as this species is now in excess of the limiting reagent, chloride ion.

$$[Ag^+] = (\text{mmol of Ag+ - mmol of Cl-})/\text{V(in mL)}$$

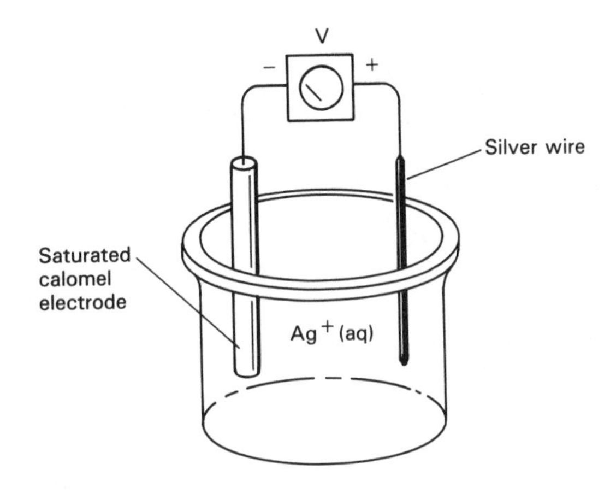

Figure 21.1 Potentiometric Titration of Chloride Ion

The two half-cell reactions and potentials that have to be considered for this experiment are

$$2\,Hg(l) + 2\,Cl^-(aq) = Hg_2Cl_2(s) + 2\,e^- \qquad E^\circ_{ox} = -\,0.268\text{ V}$$

$$Ag^+(aq) + e^- = Ag(s) \qquad E^\circ_{red} = 0.799\text{ V}$$

The concentration of the chloride ion on the inside of the standard calomel electrode that is used as a reference electrode is constant. The only concentration that changes during this experiment is the concentration of the silver(I) ion. With these considerations, the appropriate form of the Nernst Equation that describes this cell is

$$E^\circ_{cell} = 0.558 + 0.05916\log[Ag^+]$$

For the half-cell reactions as written, the concentration of silver ion in the Nernst Equation would be in the denominator but the sign of the log term would in this case be negative. The log identity, $\log x = -\log x^{-1}$, allows the function to be changed to this form.

The following symbols are used for developing the relevant equations:

VA = volume of the analyte, chloride
CA = concentration of chloride
NA = millimoles of chloride
VT = volume of titrant added, silver
CT = concentration of silver
NT = millimoles of silver

The region before the equivalence point is reached in this titration is governed by the relationship, NA > NT and

$$[Ag+] = \frac{K_{sp}}{[Cl^-]} = \frac{K_{sp} \cdot (VA + VT)}{NA - NT}$$

At the equivalence point, NA = NT and

$$[Ag+] = \sqrt{K_{sp}}$$

After the equivalence point, NA < NT and

$$[Ag+] = \frac{NT - NA}{VA + VT}$$

Worksheet 21.5 computes the cell potential as $[Ag^+]$ varies because of its addition as a titrant. Formula List 21.5 illustrates the formulas. The conditional statement in cell C14 first checks to establish if the titration is past the end-point. If NA < NT, then the computation is governed by the latter of the three equations illustrated above. If the previous inequality is false, the second conditional statement establishes if NA = NT and if true, computes the square root of the solubility product. If none of these is true, then the computation defaults to using the first equation presented above.

Formula List 21.5 Formulas for Worksheet 21.5

Cell B8	= CA*VA
Cell A14	= 0
Cell A15	= A14+10
Cell B14	= E0+0.05916*LOG10(C14)
Cell C14	= IF(NA<D14,(D14-NA)/(VA+A14), IF(NA=D14,K^.5,K*(VA+A14)/(NA-D14)))
Cell D14	= CT*A14
Cell B5	Defined as E0
Cell B6	Defined as CA
Cell B7	Defined as VA
Cell B8	Defined as NA
Cell B9	Defined as CT
Cell B10	Defined as K

	A	B	C	D
1	**Potentiometric Titration of Chloride Ion**			
2				
3	Cl- (aq) + Ag+ (aq) = AgCl (s)			
4				
5	E° (V) =	0.558		
6	[Cl-] =	0.1	analyte	
7	V(mL) Cl- =	100		
8	mmol Cl- =	10		
9	[Ag+] =	0.1	titrant	
10	Ksp =	1.8E-10		mmol
11				titrant
12	V(mL) Ag+			added
13		E(volts)	[Ag+]	Ag+
14	0	0.0407	1.8E-09	0
15	10	0.0458	2.2E-09	1
16	20	0.0511	2.7E-09	2
17	30	0.0566	3.34286E-09	3
18	40	0.0624	4.2E-09	4
19	50	0.0689	5.4E-09	5
20	60	0.0763	7.2E-09	6
21	70	0.0852	1.02E-08	7
22	80	0.0971	1.62E-08	8
23	90	0.1163	3.42E-08	9
24	100	0.2698	1.34164E-05	10
25	110	0.4206	0.004761905	11
26	120	0.4372	0.009090909	12
27	130	0.4465	0.013043478	13
28	140	0.4528	0.016666667	14

Worksheet 21.5 Potentiometric Titration of Chloride Ion

The dramatic change in the cell potential when this titration is near the endpoint is illustrated by Chart 21.5. If the cell responds in an ideal manner according to the Nernst Equation, the cell potential changes by 59 mV for each tenfold change in the silver(I) ion concentration. Potentiometric titrations are very useful as an analytical tool for establishing the concentration of an unknown. This cell responds to the changes in the silver(I) ion concentration, and it also responds to the changes in the chloride ion concentration. When used in this fashion, the cell serves as a halide electrode instead of a silver electrode. These chemical principles serve as the basis for the development of ion-selective electrodes. The most widely used ion-selective electrode is the glass electrode that measures pH. As before, the glass electrode's potential changes by 59 mV for each tenfold change in the hydrogen ion concentration.

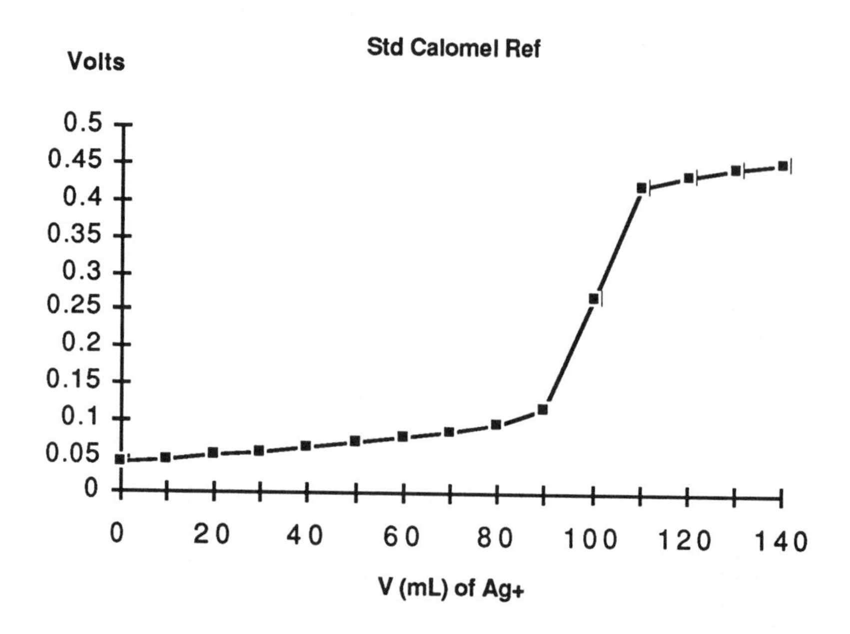

Chart 21.5 Potentiometric Titration of Chloride Ion

Problems

1. Calculate the K_{sp} for CuBr(*s*) at 298 K from the half-cell potentials:

   $Cu^{+}(aq) + e^{-} = Cu(s)$ $\quad E^{\circ}_{red} = 0.521$ V

   $CuBr(s) + e^{-} = Cu(s) + Br^{-}(aq)$ $\quad E^{\circ}_{red} = 0.030$ V

2. Calculate the K_{sp} for AgCl(*s*) at 298 K from:

   $Ag^{+}(aq) + e^{-} = Ag(s)$ $\quad E^{\circ}_{red} = 0.799$ V

   $AgCl(s) + e^{-} = Ag(s) + Cl^{-}(aq)$ $\quad E^{\circ}_{red} = 0.222$ V

3. Calculate the K_{sp} for AgI(*s*) at 298 K from:

   $Ag^{+}(aq) + e^{-} \rightarrow Ag(s)$ $\quad E^{\circ}_{red} = 0.799$ V

   $AgI(s) + e^{-} \rightarrow Ag(s) + I^{-}(aq)$ $\quad E^{\circ}_{red} = -\ 0.152$ V

4. What is the cell potential for a reaction in which one electron is transferred at 298 K if the equilibrium constant is 1.0 x 10^{-4}?

5. Use the reduction potential for reducing dicyanosilver(I) to solid silver and cyanide ion to compute the formation constant for the complex, dicyanosilver(I) at 298 K.

$$Ag(CN)_2^-(aq) + e^- = Ag(s) + 2\,CN^-(aq) \qquad E^\circ_{red} = -\,0.310\text{ V}$$

6. Calculate the equilibrium constant for producing tin(II) from tin(*s*) and tin(IV) at 298 K.

$$Sn(s) + Sn^{4+}(aq) = 2\,Sn^{2+}(aq)$$

$$Sn^{2+}(aq) + 2\,e^- = Sn(s) \qquad E^\circ_{red} = -\,0.136\text{ V}$$

$$Sn^{4+}(aq) + 2\,e^- = Sn^{2+}(aq) \qquad E^\circ_{red} = 0.154\text{ V}$$

7. Calculate the formation constant for the formation of hexafluorotin(IV) from tin(IV) and fluoride at 298 K.

$$Sn^{4+}(aq) + 6\,F^-(aq) = SnF_6^{2-}(aq)$$

$$SnF_6^{2-}(aq) + 4\,e^- = Sn(s) + 6\,F^-(aq) \qquad E^\circ_{red} = -\,0.250\text{ V}$$

$$Sn^{4+}(aq) + 4\,e^- = Sn(s) \qquad E^\circ_{red} = 0.009\text{ V}$$

8. Starting with standard state concentrations, find concentrations that produce a cell voltage of - 0.200 V at 298 K. Investigate the equilibrium region and compute the equilibrium concentrations.

$$Pb(s) + Cu^{2+}(aq) = Pb^{2+}(aq) + Cu(s)$$

$$Cu^{2+}(aq) + 2\,e^- = Cu(s) \qquad E^\circ_{red} = 0.337\text{ V}$$

$$Pb^{2+}(aq) + 2\,e^- = Pb(s) \qquad E^\circ_{red} = -\,0.126\text{ V}$$

9. Plot the potentiometric titration of chloride ion with silver(I) ion.

$[Cl^-]_0$ = 0.05 M with a volume of 80 mL

$[Ag^+]_0$ = 0.100 M

10. Calculate the cell potential for the reduction of Cr(VI) by Sn(II) in acidic solution using the following concentrations at 298 K.

	$[Cr_2O_7^{2-}]$	$[Sn^{2+}]$	$[H^+]$	$[Cr^{3+}]$	$[Sn^{4+}]$
a)	0.30 M	0.20	0.20	0.20	0.10
b)	0.15	0.05	0.70	0.01	0.03
c)	0.08	0.10	0.15	0.04	0.18

11. For these concentrations decide if the reaction is spontaneous, at equilibrium, or nonspontaneous at 298 K.

$$ClO_2^- + 2\,I^- + H_2O = ClO^- + I_2(s) + 2\,OH^-$$

$2\,I^- \rightarrow I_2(s) + 2\,e^-$ $\quad E°_{red} = -0.536$

$ClO_2^- + H_2O + 2\,e^- = ClO^- + 2\,OH^-$ $\quad E°_{red} = 0.590$

	$[ClO_2^-]$	$[I_2^-]$	$[ClO^-]$	$[OH^-]$
a)	0.100	0.100	0.0001	1.0
b)	0.010	0.010	0.010	1.0
c)	0.010	0.010	0.010	0.10
d)	0.020	0.020	0.010	0.10
e)	0.010	0.010	0.010	0.080

12. Reduction potentials of biological interest are tabulated at pH = 7, instead of unit activity for hydrogen ion. The following are standard state half-reaction reduction potentials. Compute the reduction potentials at pH = 7. They are known as *E* .

$NAD^+ + H^+ + 2\,e^- = NADH$ $\quad E°_{red} = -0.105$ V

pyruvate + $2\,H^+ + 2\,e^-$ = lactate $\quad E°_{red} = 0.224$ V

dehydroascorbate + $2\,H^+ + 2\,e^-$ = ascorbate + H_2O $\quad E°_{red} = 0.390$ V

22

Nuclear Chemistry

Chemical reactions do not change or disturb the nucleus as they alter the distribution of electrons and change the bonding between nuclei. Nuclear reactions change the nuclei themselves regardless of the chemical environment. The nuclear decay of a given isotope is independent of the chemical circumstances or physical state. All nuclear decay reactions obey first-order kinetics. Nuclear decay is described in terms of the half-life of the isotope, the percent or amount remaining, and the percent that has decayed. The forces that influence the structure of the nucleus are very large, and they play a pivotal role in determining the distribution of isotopes found on earth and in the universe. Calculation of the binding energy per nucleon for the nucleus clarifies the observed distribution of the elements and the thermodynamic instability that results in the processes of fission and fusion. Radioactive dating to determine the age of a material is an important tool in archaeology and geology. This technique is based on the first-order decay of radioactive isotopes and their half-lives. (BLB Chap. 21)

Nuclear Decay

All nuclear decay processes can be described by first-order kinetics. The equations that describe first-order kinetics were presented in Chapter 14, Chemical Kinetics. There are some 1700 radioactive isotopes known and a few of these are listed in Table 22.1. The half-lives of radioactive isotopes vary from 1×10^{-9} s to 4×10^{17} y. Carbon-containing objects that were once living that range from a few thousand to fifty thousand years old can be dated because of the radioactive isotope, carbon-14.

The half-life of an isotope can be used to compute the rate constant, percent, and amount of the isotope remaining, and the number of half-lives that have elapsed for a chosen time interval. Worksheet 22.1 illustrates the computation of the variables associated with radioactive first-order decay. The formulas and cell definitions are presented in Formula List 22.1 following Worksheet 22.1.

Table 22.1 Half-lives of Radioactive Isotopes

Isotope	**Half-life**
hydrogen-3	12.26 y
carbon-14	5730 y
phosphorus-32	14.3 d
chlorine-36	3×10^5 y
potassium-40	1.27×10^9 y
iron-59	46 d
cobalt-60	5.25 y
germanium-66	6.9 h
strontium-90	27.6 y
technetium-99	6.1 h
iodine-131	8.07 d
barium-139	85 m
uranium-238	4.5×10^9 y
plutonium-239	24,000 y
americium-241	432.2 y

	A	B	C	D
1	**Radioactive Decay**			
2				
3	**Isotope**	P-32		
4				
5	t(1/2) =	14.3	**d**	
6	t (elapsed) =	6.3	**d**	
7	Units of t	d		
8	A0 =	7.2	**g**	
9	Units of A	g		
10				
11	k =	**0.048472**	inverse	**d**
12	A =	**5.305312**	**g**	
13	t(1/2) elasped	**0.440559**		
14	% remaining	**73.68%**		

Worksheet 22.1 Radioactive Decay

If the United States now has 6×10^5 lbs of plutonium-239, compute the amount that will be present in the year 2060. What percent of the current amount will be present in 2060? The half-life of cobalt-60 is 5.25 y. How much of this isotope remains from a 1.00 mg sample in 15.9 y? What percent of the original sample remains after this period of time? How many half-lives have elapsed with this sample? What are the same values for a 1.00 mg sample of americium-241 which has a half-life of 432.2 y?

Formula List 22.1 Formulas for Worksheet 22.1

Cell B11	=LN(2)/t.5
Cell B12	=A0*EXP(-k*t)
Cell B13	=t/t.5
Cell B14	=A/A0
Cell C12	=B9
Cell D11	=B7
Cell B5	Defined as t.5
Cell B6	Defined as t
Cell B8	Defined as A0
Cell B11	Defined as k
Cell B12	Defined as A

Radioactive Dating

Radioactive dating may be used to determine the age of a sample of material. In order for the results to be correct, all of the decay product must be present in the sample. What is really measured is the time since the sample solidified. The amount of the isotope present and the amount of decay product must be converted to moles so that the original amount of the isotope (before any decay) can be calculated. Any units used for the amount of the isotope and the decay product that are the same for both are valid because the calculation involves a ratio of the moles of isotope now present compared to the moles of isotope originally present. Worksheet 22.2 is designed to calculate the age of an object from the decay of a specific radioactive isotope. The present mass of the radioactive isotope and the mass of the "sole" decay product and the half-life of the isotope must be furnished for this computation.

	A	B	C	D
1	**Dating by Radioactive Decay**			
2				
3		Original	Product	
4				
5	**Isotope**	U-238	Pb-206	
6	Amount now	1	0.257	mg
7	At Mass	238	206	
8	t(1/2) =	4500000000	y	
9				
10	Initial amount	**1.296922**	**mg**	
11	% remaining	**77.11%**		
12	t =	**1.69E+09**	**y**	

Worksheet 22.2 Radioactive Dating

Formula List 22.2 presents the formulas and cell definitions for this spreadsheet. The basis for this computation is the equation

$$t = t_{1/2} \cdot \frac{\ln (A_0/A)}{\ln 2}$$

The initial amount of the radioactive isotope that decays is calculated by assuming a one-to-one stoichiometry for the decay product and the decaying isotope after both species have been converted to moles. The total number of moles of the decaying isotope initially present is then multiplied by the atomic mass of this isotope. This calculation is accomplished in cell B10.

Table 22.2 Formulas for Worksheet 22.2

Cell B10	=B7*(C6/C7+A/B7)
Cell B11	=A/A0
Cell B12	=t.5*LN(A0/A)/LN(2)
Cell C10	=D6
Cell C12	=C8
Cell B6	Defined as A
Cell B8	Defined as t.5
Cell B10	Defined as A0

Calculate the age of a mineral sample that contains 60 mg of uranium-238 and 24 mg of lead-206. What is your conclusion about the age of the same sample if it contains 43 mg of potassium-40 and 122 mg of the decay product argon-40?

Nuclear Binding Energy

The tendency of isotopes to undergo fission and fusion is revealed by the thermodynamic stability conferred to the nucleus by the nucleons (protons and neutrons) within the nucleus. Nuclear mass can be calculated from the atomic mass by subtracting the mass of the electrons for a specific isotope. The mass of an electron is 5.486 x 10-4 amu. Table 22.2 lists the nuclear mass of a few nuclei.

The binding energy is computed from the mass defect, which is the difference between the mass calculated as the sum of the individual nuclear particles and the actual mass of the nucleus. The "apparent" loss of mass is the result of the binding energy that is needed to hold all of the nucleons (many have a positive charge) in the nucleus. The "apparent" mass loss is converted to an equivalent amount of energy by the conversion factor of 1.49×10^{-10} J/amu.

$$\Delta E = \Delta m \cdot c^2$$

For the computation of the total mass of all of the nucleons without the loss of mass for binding the nucleons, the mass of the proton and neutron are necessary. The rest mass of a proton is 1.007277 amu and that of a neutron is 1.008665 amu. Worksheet 22.3 computes the binding energy for an isotope from the nuclear mass and the numbers of protons and neutrons.

Table 22.2 Nuclear Mass (amu)

Isotope	Nuclear Mass
He-4	4.00150
Li-6	6.01347
C-12	11.9967
Fe-56	55.92066
Co-60	59.9590
Ni-60	59.9154
Ni-61	60.91570
Kr-92	91.9021
Ba-141	140.8833
U-235	234.9934
U-238	238.0003

	A	B	C
1	**Binding Energy**		
2			
3	mass nucleus =	3.01493	amu
4	# of protons =	2	
5	# of neutrons =	1	
6			
7	Mass defect =	**0.008289**	**amu**
8	Binding energy =	**1.24E-12**	**J**
9	B. E. / nucleon =	**4.12E-13**	**J**

Worksheet 22.3 Binding Energy

Compute the binding energy per nucleon for Li-6, Fe-56, and U-235. Compare the results for the three isotopes. The spreadsheet formulas are presented in Worksheet 22.3F. Cells B3, B4, B5, B7, and B8 are defined as MN, NP, NN, MD, and BE (mass of nucleus, number of protons, number of neutrons, mass defect, and binding energy).

	B
7	=(NP*1.007277+NN*1.008665)-MN
8	=0.000000000149*MD
9	=BE/(NP+NN)

Worksheet 22.3F Formulas For Worksheet 22.3

The overall trend in the binding energy per nucleon shows that iron-56 is the most stable nuclear isotope in the universe. Worksheet 22.4 is designed to compute and compare several of the binding energies per nucleon simultaneously. This spreadsheet may then be used to plot a scatter graph which illustrates the thermodynamic stability trends of the nuclei. For Chart 22.4, the scale of the binding energy per nucleon has been multiplied by one trillion, 1×10^{12}, to improve the appearance of the vertical scale. The data must be graphed as a scatter chart to accommodate missing values on the horizontal scale.

	A	B	C	D	E
1	**Binding Energy per Nucleon**				
2					
3	mass proton	1.007277	amu		
4	mass neutron	1.008665	amu		
5	convert m->E	1.49E-10	J/amu		
6					
7	Nucleons				
8		BE/nuc (J)	Mass	Protons	Neutrons
9	2	1.86E-13	2.01345	1	1
10	3	4.12E-13	3.01493	2	1
11	4	1.13E-12	4.0015	2	2
12	6	8.53E-13	6.01347	3	3
13	7	8.97E-13	7.01435	3	4
14	9	1.03E-12	9.00999	4	5
15	10	1.04E-12	10.0102	5	5
16	12	1.23E-12	11.9967	6	6
17	13	1.19E-12	13.0001	6	7
18	16	1.28E-12	15.9905	8	8
19	52	1.40E-12	51.9273	24	28
20	56	**1.41E-12**	55.9206	26	30
21	206	1.26E-12	205.9295	82	124
22	233	1.22E-12	232.989	92	141
23	234	1.22E-12	233.9904	92	142
24	235	1.21E-12	234.9934	92	143
25	238	1.21E-12	238.0003	92	146

Worksheet 22.4 Binding Energy Per Nucleon

Table 22.4 Formulas for Worksheet 22.4

Cell A9	=D9+E9
Cell B9	=mE*((D9*MP+E9*MN)-C9)/A9
Cell B3	Defined as MP
Cell B4	Defined as MN
Cell B5	Defined as mE

The scatter chart is created by selecting the range of cells from A1 to B25 on Worksheet 22.4. From the Gallery category on the menu bar select Scatter from which option 2 is selected.

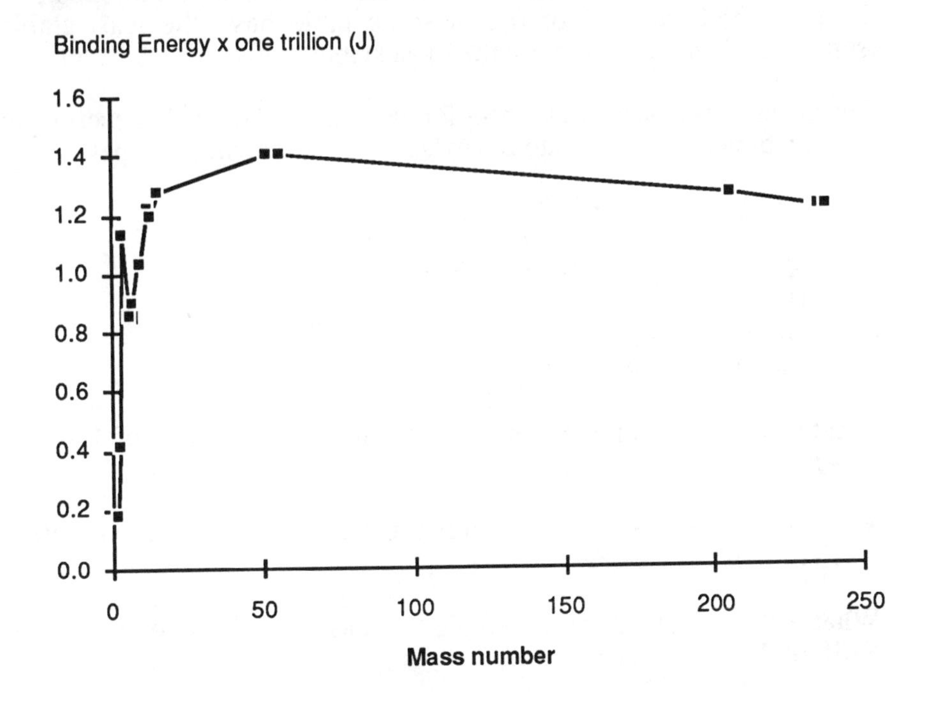

Chart 22.4 Binding Energy Per Nucleon

Problems

1. Of an initial sample of 5.7 mg, how many mg of germanium-66 remain in 6.9 hr?

2. Of an initial 87.6 mg, how many mg of hydrogen-3 remain in 10.0 yr?

3. Of an initial 24.7 mg, how many mg of barium-139 remain in 2.0 hr?

4. Of an initial sample of 0.027 mg, how many mg of technetium-99 remain in 10.0 hr?

5. How many half-lives of iron-59 have elapsed in 84 da?

6. How many half-lives of chlorine-36 have elapsed in 4.5 x 109 yr?

7. If an infant ingests strontium-90, what percent remains in the 69 year old adult?

8. Calculate the binding energy per nucleon for 6 or 7 isotopes and discuss which regions of the periodic table have the least stable isotopes and which region has the most stable.

9. Design and develop a spreadsheet for calculating the binding energy per nucleon directly from the atomic mass for the following isotopes:

   a) Ni-61 60.93106 amu
   b) Pb-206 205.97447
   c) H-3 3.01605
   d) He-3 3.01603
   e) U-235 235.0439
   f) Ba-141 140.9140

10. What is the age of a mineral sample that has a weight ratio of Ar-40 to K-40 of 1.7?

11. What is the age of a mineral sample that has a weight ratio of Ar-40 to K-40 of 0.59?

12. What is the age of a mineral sample that has a weight ratio of Ar-40 to K-40 of 2.9?

13. What is the age of a mineral sample that contains 119 mg of U-238 and 103 mg of Pb-206?

14. What is the age of a mineral sample that contains 68 mg of U-238 and 19.6 mg of Pb-206?

23

Experimental Data

The Beer-Lambert law for the absorption of light provides the chemist with a convenient method for determining the concentration of the absorbing species. A Beer's law graph is developed from several solutions of known concentration. From this plot, the molar absorptivity can be determined for use in analyzing solutions of unknown concentration. A spreadsheet for interpreting spectrophotometric data is presented. The use of scatter charts for illustrating the linear least-squares values and the experimental data points is explored for spectrophotometric and kinetic data. Both first- and second-order kinetic data are interpreted using these methods. The interpretation and verification of kinetic data is approached from five different perspectives. Verification that the data represents first- or second-order kinetic data is accomplished by predicting the molar concentration or absorbance from the initial data measurement and the time units associated with the experimental data. The value of a rate constant, k, is determined from each data point as well as from the linear least-squares technique. A slope between consecutive data points is computed. An average rate constant for consecutive data points is computed from the average rate divided by the average concentration over that interval. And last, the predicted linear least-squares line is compared with the experimental data points on a scatter chart for a visual verification of the applicability of the mathematical treatment of the kinetic data. The mathematical treatment of spectrophotometric first-order kinetic data is presented. (BLB Chap. 14)

Spectrophotometry

When a chemical species absorbs a photon of electromagnetic radiation, the energy of the species is increased and the radiant power of the light beam is decreased. Spectrophotometers measure the decrease in radiant power as a function of wavelength. It is common to speak of a specific wavelength as though the instrument has the capability of sorting photons to a single selected energy, but in truth, all instruments use a diffraction grating or prism which selects a distribution of wavelengths. The distribution is usually described by the Gaussian function that was presented in Chapter 3. Spectrophotometers measure either the percent transmittance or the absorbance of a substance at a selected wavelength. Absorbance is a more useful quantity for chemists because the concentration of the chemical species that causes the decrease in the radiant power of the light beam can be calculated from the Beer-Lambert law.

$$A_\lambda = \varepsilon_\lambda \cdot l \cdot c$$

The absorbance, A, is a function of wavelength and it is dimensionless. The molar absorptivity, ε, is a function of wavelength and it is expressed in units of $M^{-1}cm^{-1}$. The path length, l, is expressed in centimeters and the concentration, c, in moles per liter (molar). For analytical purposes, the absorption spectrum of a material is first obtained over a selected range of the visible or ultraviolet regions of the electromagnetic spectrum. An analysis for the concentration of a species is most sensitive when an absorption maximum is chosen. When a wavelength has been chosen for the analysis, four or five solutions that have different concentrations of the substance are measured at the chosen wavelength with a spectrophotometer.

Worksheet 23.1 computes the molar absorptivity and its standard deviation from experimental data. The data on the concentrations of the known solutions are entered in the first column. The experimental absorbances of the changing concentrations are entered in the second column. The LINEST function computes the slope and y-intercept of the line of regression for the two random variables, concentration and absorbance. The LINEST function uses the method of least-squares for determining the best fit for the data. The slope determined by the linear least-squares method is the molar absorptivity of the species. The computed molar absorptivity is used to calculate the predicted absorbance in the third column. This column serves to either validate or vitiate that Beer's law is valid over the chosen range of concentrations. All of the computations of the predicted absorbance can be shown, or, if a scatter graph of these results is desired, the intermediate values can be replaced with #N/A (No value is Available). With line and scatter charts, the value, #N/A, causes the previous point and the next point to be joined, skipping the #N/A point or points. The standard deviation is computed using an array formula with the STDEV function. It should be noted that this is not the correct standard deviation for the molar absorptivity as determined by the method of linear least-squares. The standard deviation computed in this manner is associated with the average value of the molar absorptivity as calculated from the last four data points. This average is 5122 instead of the linear least-squares value of 4797. Most of the difference is a reflection of the nonzero value for the absorbance at zero concentration of the species of interest.

	A	B	C	D	E
1	**Beer-Lambert Law**		Fe(SCN)++		
2				LINEST	Least Sq
3				slope	intercept
4	[Fe(SCN)++]	Experimental	Predicted	4797	0.04
5		Absorbance	Absorbance		
6	0	0.04	0.04		
7	0.00008	0.451	#N/A		
8	0.00016	0.8	#N/A		
9	0.0002	0.97	#N/A		
10	0.0003	1.5	1.48		
11					
12	M absorptivity	4797			
13	std deviation	351			

Worksheet 23.1 Determining Molar Absorptivity

Worksheet 23.1F illustrates the formulas used in Worksheet 23.1. The LINEST function returns, as a horizontal array of two elements, the slope and y-intercept of the line of regression, $y = mx + b$, for the two random variables, x and y, represented by measured x's and measured y's. Before entering this function, select two cells that are adjacent in a horizontal direction and then type the function in the left cell. Press the Shift + Ctrl + Enter (IBM®) or Command + Return (Macintosh®) in order to enter the contents in both cells. If you wish to remove these formulas, be sure that you select both cells before using the Clear command under Edit on the menu bar. Then select Formulas and OK to finish removing the LINEST function. The array formula used for computing the standard deviation is also entered by pressing Shift + Ctrl + Enter (IBM®) or Command + Return (Macintosh®).

	C	D	E
4	Predicted	=LINEST(B6:B10,A6:A10)	=LINEST(B6:B10,A6:A10)
5	Absorbance		
6	=m*A6+b		
7	#N/A		
8	#N/A		
9	#N/A		
10	=m*A10+b		

	B
12	=D4
13	=STDEV(B7:B10/A7:A10)

Worksheet 23.1F Formulas For Worksheet 23.1

Formula List 23.1 Defined Variables For Worksheet 23.1

Cell D4	Defined as m
Cell E4	Defined as b

A scatter chart of the experimental data and the linear least-squares fit of this data is shown as Chart 23.1. On Worksheet 23.1, select cells A5 through C10. Then choose File, New, and double click on Chart. With the appearance of a chart, select Gallery, Scatter, and choose option 1. A new scatter graph will appear and then select Chart and Add Overlay. At this point, the linear least-squares line is incorrect. Choose Format, Overlay, and double click on Scatter. The chart is now complete.

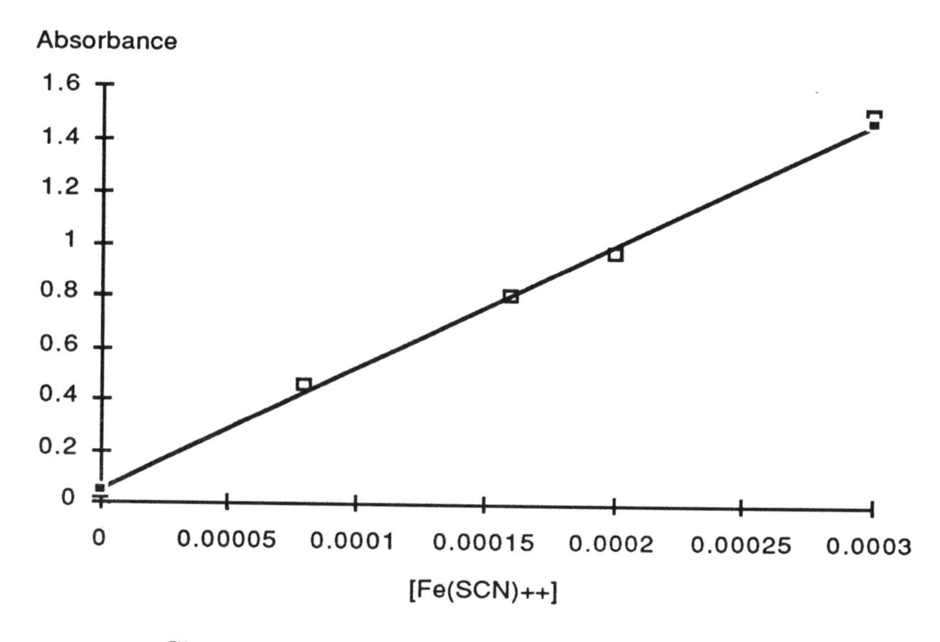

Chart 23.1 Determination of the Molar Absorptivity

First-Order Kinetics

Chapter 14, Chemical Kinetics, presents the principles and equations needed for the interpretation of first-order kinetic data. Worksheet 23.2 illustrates the design of a spreadsheet for verifying that a set of data represents first-order kinetics. The experimental data are for the rearrangement of methyl isonitrile, CH_3NC in the gas phase at 488 K. The two first-order kinetic equations that are needed for this spreadsheet are:

$$[A]_t = [A]_{t=0} e^{-kt}$$

$$k = (1/t) \cdot \ln([A]_0/[A]_t)$$

The experimental values for time and concentration are entered in columns one and two. The third column computes the predicted concentrations from the value of *k* which is determined by the linear least-squares method with the assumption that the data represents first-order kinetics. The fourth column computes the first-order rate constant from the initial concentration, $B$6, and the pair of data points for concentration and time in that row. The fifth column computes the slope between the previous point and the current point for a graph that has the ln $[CH_3NC]_t$ as the vertical axis and t as the horizontal axis. The sixth column computes the average rate divided by the average concentration raised to the first power which is equal to the first-order rate constant. The lower group of columns is for graphing the results and for computing the linear least-squares slope and intercept. The column of cells, B15 through B20, computes the ln $[CH_3NC]_t$ for use in determining the "best fit value" for the slope. The LINEST function uses the natural log of concentration, ln $[CH_3NC]_t$, as the y's and time as the x's. The slope of the line of regression for the two experimental variables is the negative value of *k*.

	A	B	C	D	E	F
1	First-Order Kinetics			LINEST	Least Sq	
2				slope	intercept	
3				-0.000207	-4.099415	
4	t (m)		predicted			average
5		[CH3NC]	[CH3NC]	k	slope	rate/conc
6	0	0.0165	0.01650			
7	2000	0.011	0.01090	0.000203	-0.000203	-0.000200
8	5000	0.00591	0.00585	0.000205	-0.000207	-0.000201
9	8000	0.00314	0.00314	0.000207	-0.000211	-0.000204
10	12000	0.00137	0.00137	0.000207	-0.000207	-0.000196
11	15000	0.00074	0.00073	0.000207	-0.000205	-0.000199
12						
13	t (m)		predicted		average =	-0.0002
14		ln [CH3NC]	ln [CH3NC]			
15	0	-4.1	-4.1			
16	2000	-4.5	#N/A			
17	5000	-5.1	#N/A			
18	8000	-5.8	#N/A			
19	12000	-6.6	#N/A			
20	15000	-7.2	-7.2			

Worksheet 23.2 First-Order Kinetics

	B	C	D
6	0.0165	=B6	
7	0.011	=$B$6*EXP($D$3*A7)	=LN($B$6/B7)/A7
8	0.00591	=$B$6*EXP($D$3*A8)	=LN($B$6/B8)/A8
9	0.00314	=$B$6*EXP($D$3*A9)	=LN($B$6/B9)/A9
10	0.00137	=$B$6*EXP($D$3*A10)	=LN($B$6/B10)/A10
11	0.00074	=$B$6*EXP($D$3*A11)	=LN($B$6/B11)/A11
12			
13		predicted	
14	ln [CH3NC]	ln [CH3NC]	
15	=LN(B6)	=LN(C6)	
16	=LN(B7)	#N/A	
17	=LN(B8)	#N/A	
18	=LN(B9)	#N/A	
19	=LN(B10)	#N/A	
20	=LN(B11)	=LN(C11)	

	E	F
3	=LINEST(B15:B20,A6:A11)	
4		
5	slope	av rate/conc
6		
7	=(LN(B7)-LN(B6))/(A7-A6)	=2*(B7-B6)/((A7-A6)*(B6+B7))
8	=(LN(B8)-LN(B7))/(A8-A7)	=2*(B8-B7)/((A8-A7)*(B7+B8))
9	=(LN(B9)-LN(B8))/(A9-A8)	=2*(B9-B8)/((A9-A8)*(B8+B9))
10	=(LN(B10)-LN(B9))/(A10-A9)	=2*(B10-B9)/((A10-A9)*(B9+B10))
11	=(LN(B11)-LN(B10))/(A11-A10)	=2*(B11-B10)/((A11-A10)*(B10+B11))
12		
13	average =	=AVERAGE(F7:F11)

Worksheet 23.2F Formulas for Worksheet 23.2

Both cells D3 and E4 contain the linear least-squares regression formula, =LINEST(B15:B20,A6:A11), for evaluating the slope and intercept of the line of regression for the first-order kinetics plot. The vertical axis contains the natural log of the concentration and the horizontal axis contains the time variable. When the natural log of the concentration is plotted, the slope of the least-squares fit is the negative of the first-order rate constant, *k*. When the base-ten log is plotted, the slope multiplied by the natural log of ten represents the negative of the first-order rate constant.

Chart 23.2 represents a plot of the natural log of the concentration versus time. All values for the concentration are less than one with the result that the natural log of these values is negative. As the concentration becomes smaller, the natural log of the concentration becomes more negative.

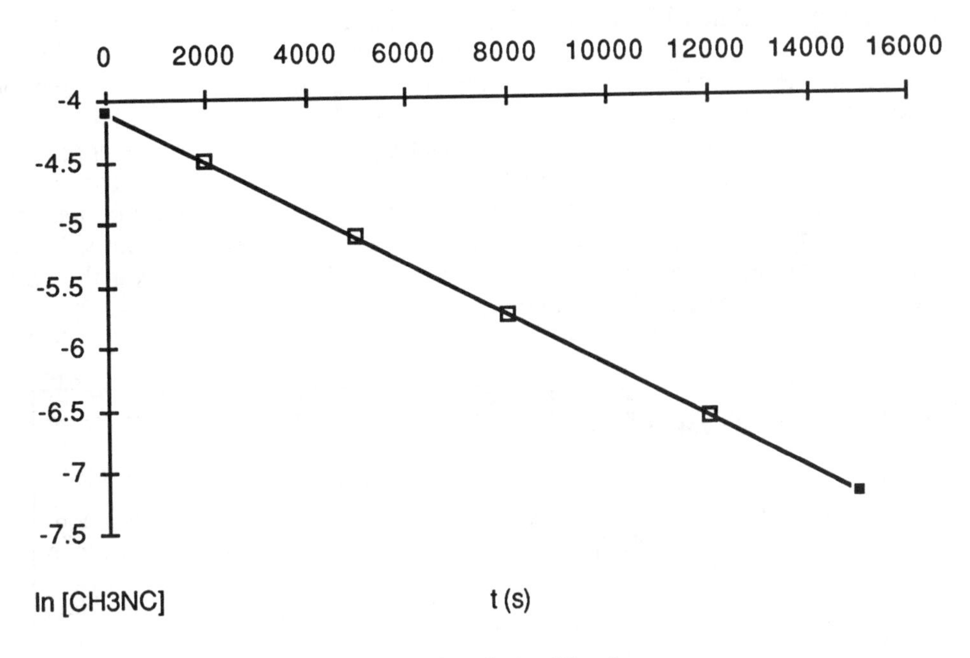

Chart 23.2 First-Order Kinetics

Second-Order Kinetics

Worksheet 23.3 presents a spreadsheet for verifying that a set of data represents second-order kinetics. The experimental data is for the gas phase decomposition of nitrogen dioxide at 320°C.

$$2\ NO_2\ (g) = 2\ NO\ (g) + O_2\ (g)$$

The two second-order kinetic equations necessary for this spreadsheet are:

$$[A]_t = (k \cdot t + 1/[A]_0)^{-1}$$

$$k = 1/t \cdot (1/[A]_t - 1/[A]_0)$$

Experimental data for time and concentration are entered in columns one and two. The third column computes the predicted concentrations from the value of k as determined by the linear least-squares technique with the assumption that the data represents second-order kinetics. Column four computes the second-order rate constant from the initial concentration, $B$6, and the individual data points for concentration and time. The fifth column computes the slope between the previous point and the current point for a graph that has 1/$[NO_2]_t$ as the vertical axis and t as the horizontal axis. The sixth column computes the second-order rate constant, which is equal to the average rate divided by the average concentration raised to the second power. The lower group of columns is used for producing a chart of the results and for computing the linear least-squares slope and intercept. The column of cells, B17 through B24, computes 1/$[NO_2]_t$ for use in determining the "best fit value" of the slope. The LINEST function uses the inverse of the concentration, 1/$[NO_2]_t$, as the y's and time as the x's. The slope of the line of regression for the two experimental variables is the value of the second-order rate constant, *k*.

	A	B	C	D	E	F
1	**Second-Order Kinetics**			LINEST	Least Sq	
2				slope	intercept	
3				1.0038	9.86	
4	t (s)		predicted			average
5		[NO2]	[NO2]	k	slope	rate/conc
6	0	0.1000	0.1000			
7	10	0.0506	0.0499	0.98	0.98	-0.87
8	20	0.0329	0.0332	1.02	1.06	-1.02
9	30	0.0253	0.0249	0.98	0.91	-0.90
10	40	0.0204	0.0199	0.98	0.95	-0.94
11	50	0.0163	0.0166	1.03	1.23	-1.22
12	60	0.0143	0.0142	1.00	0.86	-0.85
13	70	0.0125	0.0125	1.00	1.01	-1.00
14						
15	t (s)		predicted		average =	-0.971
16		1/ [NO2]	1/ [NO2]			
17	0	10.0	10.0			
18	10	19.8	#N/A			
19	20	30.4	#N/A			
20	30	39.5	#N/A			
21	40	49.0	#N/A			
22	50	61.3	#N/A			
23	60	69.9	#N/A			
24	70	80.0	80.3			

Worksheet 23.3 Second-Order Kinetics

Cells D3 and E4 contain the formula, =LINEST(B17:B24,A6:A13), the linear least-squares regression formula, for evaluating the slope and intercept of the second-order plot. A plot of 1/[concentration] versus *t* is linear, with a slope equal to the second-order rate constant. Cell D3 computes the slope of the linear least-squares regression line which best fits this data and this value is the second-order rate constant.

	B	C	D
6	0.1	=B6	
7	0.0506	=($D$3*A7+1/$B$6)^-1	=(1/B7-1/$B$6)/A7
8	0.0329	=($D$3*A8+1/$B$6)^-1	=(1/B8-1/$B$6)/A8
9	0.0253	=($D$3*A9+1/$B$6)^-1	=(1/B9-1/$B$6)/A9
10	0.0204	=($D$3*A10+1/$B$6)^-1	=(1/B10-1/$B$6)/A10
11	0.0163	=($D$3*A11+1/$B$6)^-1	=(1/B11-1/$B$6)/A11
12	0.0143	=($D$3*A12+1/$B$6)^-1	=(1/B12-1/$B$6)/A12
13	0.0125	=($D$3*A13+1/$B$6)^-1	=(1/B13-1/$B$6)/A13
14			
15		predicted	
16	1/ [NO2]	1/ [NO2]	
17	=1/B6	=1/C6	
18	=1/B7	#N/A	
19	=1/B8	#N/A	
20	=1/B9	#N/A	
21	=1/B10	#N/A	
22	=1/B11	#N/A	
23	=1/B12	#N/A	
24	=1/B13	=1/C13	

	E	F
3	=LINEST(B17:B24,A6:A13)	
4		
5	slope	av rate/conc
6		
7	=(1/B7-1/B6)/(A7-A6)	=4*(B7-B6)/((A7-A6)*(B6+B7)^2)
8	=(1/B8-1/B7)/(A8-A7)	=4*(B8-B7)/((A8-A7)*(B7+B8)^2)
9	=(1/B9-1/B8)/(A9-A8)	=4*(B9-B8)/((A9-A8)*(B8+B9)^2)
10	=(1/B10-1/B9)/(A10-A9)	=4*(B10-B9)/((A10-A9)*(B9+B10)^2)
11	=(1/B11-1/B10)/(A11-A10)	=4*(B11-B10)/((A11-A10)*(B10+B11)^2)
12	=(1/B12-1/B11)/(A12-A11)	=4*(B12-B11)/((A12-A11)*(B11+B12)^2)
13	=(1/B13-1/B12)/(A13-A12)	=4*(B13-B12)/((A13-A12)*(B12+B13)^2)
14		
15	average =	=AVERAGE(F7:F13)

Worksheet 23.3F Formulas for Worksheet 23.3

Chart 23.3 represents a scatter chart of the experimental data and the linear least-squares fit. The vertical axis contains the reciprocal of the concentration of nitrogen dioxide and the horizontal axis contains the time variable. Detailed information on charts is presented in Appendix C. The charting options under Gallery on the menu bar have two important options for plotting chemical data. The Line... option is appropriate when the data that are to be plotted on the category axis (horizontal axis) are linear and the chosen intervals represent equal values. If the category axis is linear but the intervals are not equal, then the Scatter... option is appropriate. Chart 23.3 is a scatter chart produced by executing the following steps. The three columns, A, B, and C are selected from rows 17 through 24. The following steps are then executed in sequence: File, New, Chart, Gallery, Scatter..., 1, Chart, Add Overlay, Format, Overlay..., Scatter. The text for the horizontal and vertical axes is then added in the manner described in Appendix C. In this case, a line chart is also appropriate because of the equal intervals on the horizontal axis. The following sequence of operations will produce a similar chart: File, New, Chart, Gallery, Line..., 3, Chart, Add Overlay.

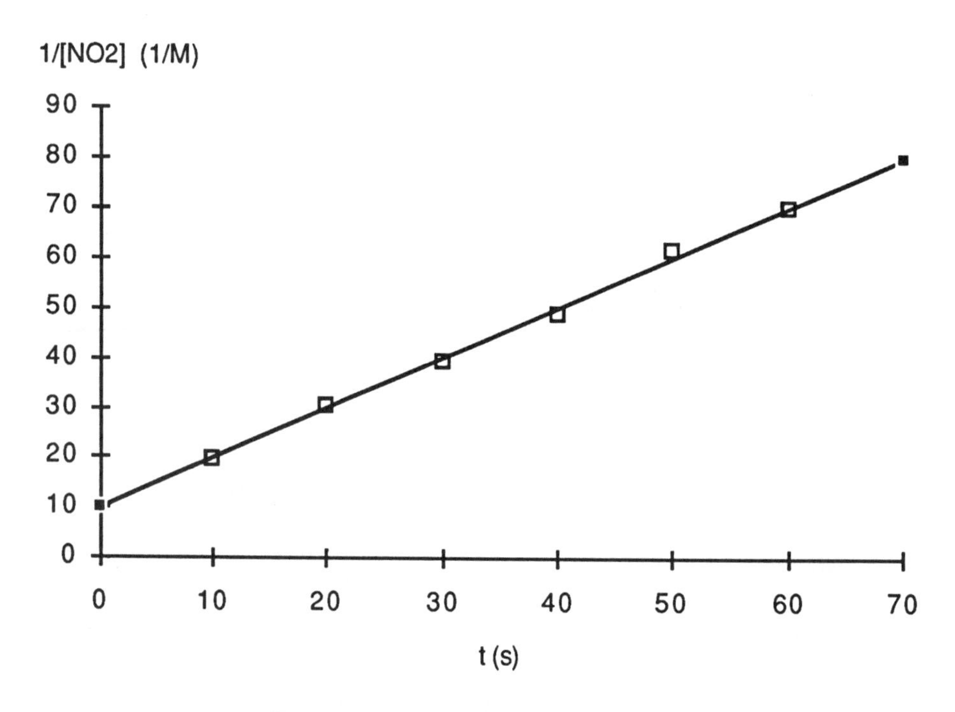

Chart 23.3 Second-Order Kinetics

Spectrophotometric Kinetic Data

The analysis and interpretation of spectrophotometric kinetic data is similar to that of concentration data. For simplicity, this discussion will assume that the reacting species has a molar absorptivity much larger than any of the other species. Beer's law is a linear relationship that establishes that the absorbance of a chemical species is directly proportional to the concentration. The proportionality constant is the product of the path length and the molar absorptivity at a selected wavelength. The change in the absorbance, $|A_{t=0} - A_{t=\infty}|$, during a chemical reaction is proportional to the change in the concentration of the absorbing species. The change in the absorbance at any time, *t*, divided by the total change in absorbance during the reaction is the extent of reaction. This may be expressed as

$$\text{extent of reaction} = \frac{|A_{t=0} - A_t|}{|A_{t=0} - A_{t=\infty}|}$$

This is a general expression that is valid regardless of the kinetic order of the reaction.

If the kinetic data is first-order, the following equations govern the behavior of the system.

$$[A]_t = [A]_{t=0}e^{-kt}$$

$$A_{\lambda,\tau} = \varepsilon_\lambda \cdot l \cdot [A]_t$$

Worksheet 23.4 represents spectrophotometric data that is generated by our design and then interpreted. The validity of the technique is demonstrated by the final outcome, which establishes the rate constant as 0.46 s^{-1}, exactly the value that was chosen to generate the absorbance data. The data are generated using first-order kinetics with an initial concentration of 0.100 M and an assumed molar absorptivity of 10.0 M^{-1} cm^{-1} for the reacting species. The important aspect of the interpretation is the necessity of subtracting the absorbance at $t = \infty$ from all of the absorbance values and then computing the natural log of each. On the spreadsheet, this quantity is represented symbolically by ln(At - Ai), when Ai represents the absorbance at infinite time.

	B	C	D
3		=LINEST(B5:B19,A5:A19)	=LINEST(B5:B19,A5:A19)
4	ln(At-Ai)	Absorbance	extent react
5	=LN(C5-$C$27)	=EXP(-0.46*A5)+0.4	=(C5-$C$22)/($C$5-$C$22)
6	=LN(C6-$C$27)	=EXP(-0.46*A6)+0.4	=(C6-$C$22)/($C$5-$C$22)

Worksheet 23.4F Formulas for Worksheet 23.4

Column C generates the absorbance data as though it were produced by an experiment. The rate constant is arbitrarily chosen as 0.46 s^{-1} and with a path length of 1.0 cm, the molar absorptivity and initial concentration of the reactant combine to produce a constant of unity. Column B computes the natural log of absorbance at *t* minus the absorbance at infinite time. The linear least-squares technique computes the best fit of this data, which does not have the inherent errors of experimental data.

	A	B	C	D
1	First-Order Kinetics		LINEST	Least Sq
2			slope	intercept
3	t		-0.461	0.00455
4		ln(At-Ai)	Absorbance	extent react
5	0	0.000	1.400	100%
6	1	-0.460	1.031	63%
7	2	-0.920	0.799	40%
8	3	-1.380	0.652	25%
9	4	-1.840	0.559	16%
10	5	-2.300	0.500	10%
11	6	-2.761	0.463	6%
12	7	-3.221	0.440	4%
13	8	-3.682	0.425	2%
14	9	-4.143	0.416	2%
15	10	-4.604	0.410	1%
16	11	-5.066	0.406	1%
17	12	-5.530	0.404	0%
18	13	-5.996	0.403	0%
19	14	-6.466	0.402	0%
20	15	-6.941	0.401	0%
21	16	-7.425	0.401	0%
22	17	-7.926	0.400	0%
23	18	-8.453	0.400	0%
24	19	-9.030	0.400	0%
25	20	-9.708	0.400	0%
26	21	-10.658	0.400	0%
27	22		0.400	0%

Worksheet 23.4 First-Order Kinetic Absorbance Data

Real experimental data have an inherent random error that is not present in this illustration. The chart of this data was obtained by following this sequence of steps. From the menu bar, File was selected, then New..., followed by Chart. Then Gallery, Line..., 1 were selected, and the cursor was clicked on the vertical axis. This allows the scale and intersection of the two axes to be adjusted to our specifications. Select Format, Scale, and enter "-7" for Category Axis Crosses at, "OK". The text for the chart is added by the following sequence: Chart, Select Chart, "ln(At-Ai)", "Enter", place the cursor on the typed label, press the mouse button and move, "Enter", and "click the mouse". For the horizontal axis, select Chart, Attach Text, Category Axis, "t (m)", "Enter", and "click the mouse". A chart title can be added with Chart, Attach Text, and Chart Title.

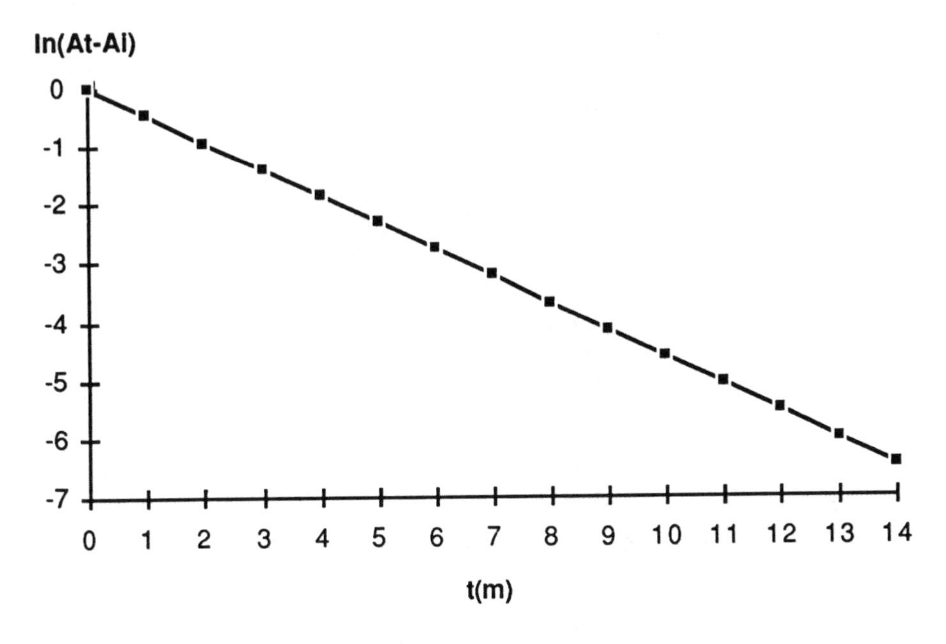

Chart 23.4 Plot of First-Order Absorbance Data

Problems

1. Compute the appropriate rate constant for the following data on the decomposition of trichloroacetate in acidic aqueous solution at 80°C. What will the concentration of the anion be when 5000 s have elapsed? What is the concentration of trichloroacetate after 1200 s? When will the extent of reaction be greater than 99.9%? At what time will the extent of reaction be greater than 25.0%?

$$H^+ (aq) + CCl_3COO^- (aq) = CO_2 (aq) + CHCl_3 (aq)$$

time (s)	$[CCl_3COO^-]$
0	0.0729
6000	0.0457
12000	0.0291
18000	0.0182
24000	0.0118
30000	0.00734
36000	0.00473
36000	0.00295

2. Compute the appropriate rate constant for the following data on the decomposition of trichloroacetate in acidic aqueous solution at 90°C.

$$H^+ (aq) + CCl_3COO^- (aq) = CO_2 (aq) + CHCl_3 (aq)$$

time (s)	$[CCl_3COO^-]$
0	0.0435
2000	0.0227
4000	0.0128
6000	0.00671
8000	0.00359
10000	0.00197
15000	0.000404

3. Compute the appropriate rate constant for the following data on the decomposition of trichloroacetate in acidic aqueous solution at 70°C.

$$H^+ (aq) + CCl_3COO^- (aq) = CO_2 (aq) + CHCl_3 (aq)$$

time (s)	$[CCl_3COO^-]$
0	0.0725
20000	0.0520
40000	0.0360
60000	0.0260
80000	0.0186
100000	0.0130
300000	0.00043

4. Using data from the three previous problems, compute E_a for the decomposition of trichloroacetate ion in acidic aqueous solution over the temperature range of 70°C to 90°C.

$$\ln k = -\frac{E_a}{R} \cdot \left\{\frac{1}{T}\right\} + \ln A$$

5. Using the equation present in problem 4, compute E_a for the decomposition of nitrogen dioxide.

$$2\,NO_2\,(g) = 2\,NO\,(g) + O_2\,(g)$$

t (°C)	k ($M^{-1}\,s^{-1}$)
320 °C	1.00 $M^{-1}\,s^{-1}$
330 °C	1.47 $M^{-1}\,s^{-1}$
340 °C	2.12 $M^{-1}\,s^{-1}$
350 °C	3.04 $M^{-1}\,s^{-1}$

6. Using the equation present in problem 4, compute E_a for the decomposition of hydrogen iodide.

$$2\,HI\,(g) = H_2\,(g) + I_2\,(g)$$

t (°C)	k ($M^{-1}\,s^{-1}$)
440 °C	$2.69 \times 10^{-3}\ M^{-1}\,s^{-1}$
460 °C	$6.21 \times 10^{-3}\ M^{-1}\,s^{-1}$
480 °C	$1.40 \times 10^{-2}\ M^{-1}\,s^{-1}$
500 °C	$3.93 \times 10^{-2}\ M^{-1}\,s^{-1}$

7. Radiation data for the decay of lead-210, itself a decay product of radon-222, yielded the following results. Compute the half-life of this isotope in years using the linear least-squares technique.

$$^{210}_{82}Pb = {}^{210}_{83}Bi + {}^{0}_{-1}e$$

time (d)	counts/min
0	888
100	880.1
200	873.2
300	865.2
400	858.1
500	850.2
600	843.1
700	835.9

8. The decay of tritium, H-3, yielded the following radiation data. Compute the least-squares half-life of this isotope in years.

$$^{3}_{1}H = {}^{3}_{2}He + {}^{0}_{-1}e$$

time (d)	counts/min
0	2201
100	2176
200	2117
300	2108
400	2074
500	2028
600	2008
700	1977

9. The total chromium ion (II, III, and VI) present in a solution can be analyzed by a spectrophotometric method as chromate ion after treatment with NaOH and H_2O_2. From the following data on known solutions obtained at 372 nm, determine the linear least-square molar absorptivity of chromate(VI) ion.

Concentration	Absorbance
0	0.003
8.0×10^{-5}	0.383
1.6×10^{-4}	0.774
2.0×10^{-4}	0.961
3.0×10^{-4}	1.452

A

Using Microsoft® Excel

Microsoft® Excel is an integrated spreadsheet, database, and graphics software package. It is a powerful computational tool that extends your ability to perform a variety of tasks for science and engineering. You can

- Develop worksheets to store, compute, and manipulate data
- Create charts of seven different types, including line and scatter
- Establish and analyze databases
- Create your own macro designed for a specific task

Microsoft® Excel Application and Worksheet Window

Figure A.1 illustrates the important regions of the application and worksheet window. The Menu bar contains the major selections for manipulating information contained on a spreadsheet. This is the second band from the top of the figure that contains the items: File, Edit, Formula, Format, Data, Options, Macro, and Window. The Formula bar, which is empty in the figure, will contain the formula or alphanumeric characters that you enter or have entered in a cell. This is the rectangle on the far right of the third band of the figure. The rectangle on the far left of the third band identifies the cell, A1, that is currently active. The status bar is on the left at the bottom of Figure A.1. The hyphen in the upper left corner contains the Application Control Menu, which is not useful for most of our applications (except for closing windows). The arrows in the upper right corner control the size of the application window and are of little use for our needs. The single item, File, on the menu bar is very useful. Clicking on File produces six options.

- New... creates a new, empty worksheet
- Open... opens a worksheet that is stored on the disk
- Delete... deletes a file that is stored on the disk
- Exit leaves Excel

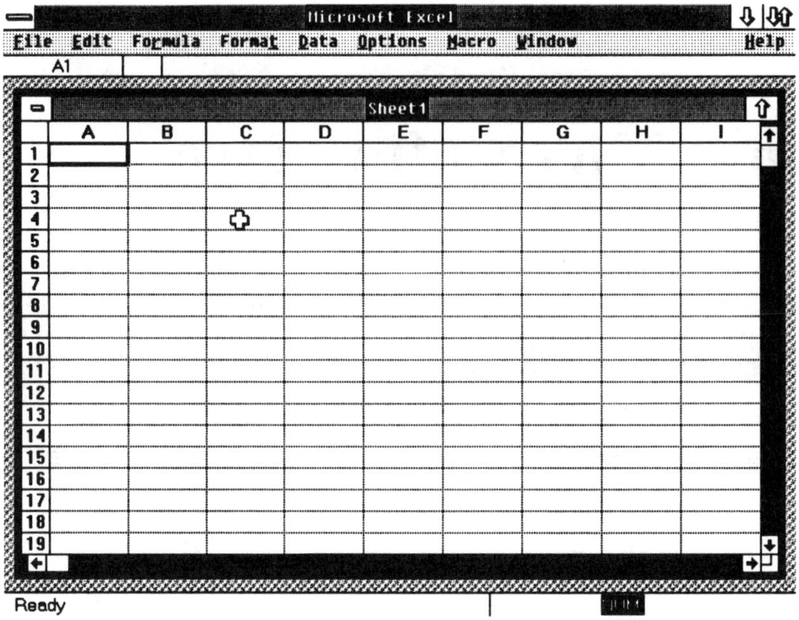

Figure A.1 Microsoft® Excel's Application and Worksheet Window

The worksheet has a narrow dark border that distinguishes it from the application window. The worksheet has a hyphen, title bar, and arrow at the top. The bottom of the worksheet contains the horizontal scroll arrows and box, and the right side contains the vertical scroll arrows and box. The remainder of the worksheet is partitioned into cells with their column and row designations at the top and left side respectively. Scrolling through the cells is accomplished by moving the box or clicking on an arrow.

Microsoft® Excel Worksheet Window

Figure A.2 shows an empty worksheet and a number of the special features. Until you save your worksheet, the title bar will display the name *Sheet* 1. The worksheet is divided into a grid of columns (letters) and rows (numbers). The hyphen in the upper left corner contains the Document Control Menu, which is similar to the Application Control Menu. This menu contains six commands: Restore, Move, Size, Maximize, Close, and Split. The arrow in the upper right corner expands the size of the worksheet. The Restore command reduces the worksheet to its previous size. The subtle dark bar at the top of the right vertical border allows you to split the single window into two separate window panes.

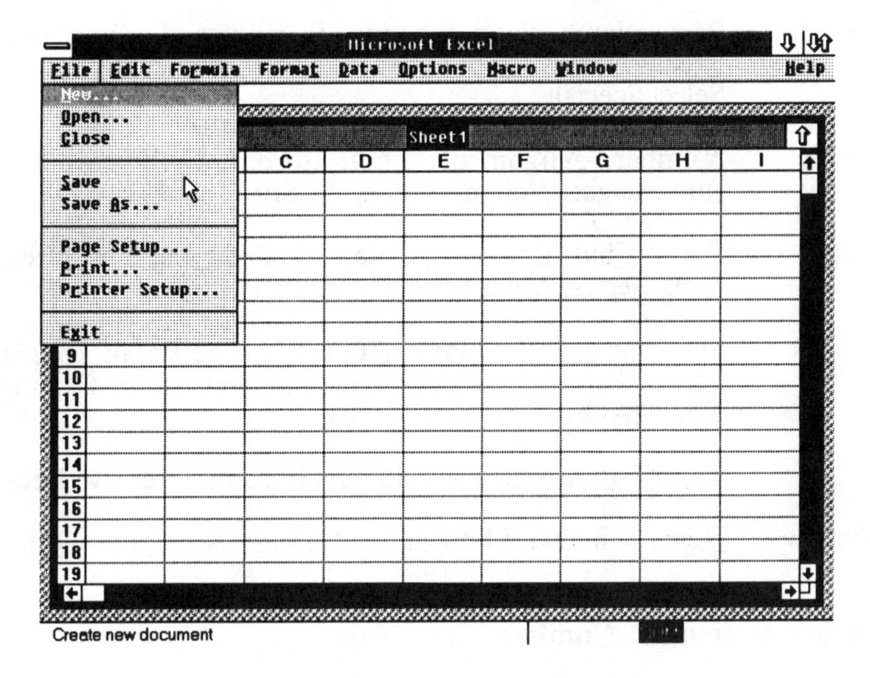

Figure A.2 Microsoft® Excel's Worksheet Window

- Restore restores the window to its previous size
- Move allows moving of the window without changing its size
- Size allows enlargement or reduction of the size of the window
- Maximize enlarges the window to its maximum size
- Close closes the active document window
- Split allows dividing of, a window into panes

Placing the cursor in a cell with the mouse or directional arrows on the keyboard allows you to enter information in the cell. Figure A.2 illustrates the drop-down menu under File on the menu bar. To activate a specific drop-down menu, place the cursor on the selection and click the mouse. The following is a summary of the various commands that are available for each of the categories on the menu bar.

- File New, Open, Close, Save, Save As, Page Setup, Print, Printer Setup, Exit
- Edit Undo Delete, Cut, Copy, Paste, Clear, Delete, Insert, Fill right, Fill Down

- Formula — Paste Name, Paste function, Reference, Define Name, Create Names, Apply Names, Note, Goto, Find, Replace, Select Special
- Format — Number, Alignment, Font, Border, Cell Protection, Row Height, Column Width, Justify
- Data — Form, Find, Extract, Delete, Set Database, Set Criteria, Sort, Series, Table, Parse
- Options — Set Print Area, Set Print Titles, Set Page Break, Display, Freeze Panes, Protect Document, Calculation, Calculate Now, Workspace, Short Menus
- Macro — Record, Run, Start Recorder, Set Recorder, Relative Record
- Window — New Window, Show Info, Arrange All, Hide, Unhide

Entering Formulas, Numbers, and Text

The syntax for the entry of a formula in a cell must always begin with an equal sign. Each value or cell that represents a value must be joined by an operator. These operators have a hierarchy that is an important consideration when determining whether sets of parentheses are needed to ensure the proper order of computation. Table A.1 illustrates the priorities of the operators available with Excel. If you are in doubt you may always include the parentheses as a safeguard.

Table A.1 Hierarchy of Operators

:	Range
space	Intersection
,	Union
-	Negation
%	Percent
^	Exponentiation
* and /	Multiplication and Division
+ and -	Addition and Subtraction
&	Join Text
< = > <= >= <>	Comparison

When information is entered into a cell, it is displayed in the formula bar. To the left of the formula bar is a check box which is an enter command that is activated when the cursor is clicked in this box. The x box is the cancel command which will cancel an entry within a cell or cancel an edit. The cell and formula bar are active until Enter or the directional arrows are keyed. The Paste Function command under Formula on the menu bar allows you to enter a mathematical formula and the terms for its argument. You may enter these functions from scratch. The Paste Function command is a quick technique for finding all of the mathematical functions that excel supports and their unique syntax. After a formula has been entered into a cell, it can be copied into other adjacent cells in the selected row or column, either to the left or right, or up or down. These commands are present on the Edit drop-down menu.

- Fill Down — select edit, cells filled adjacent and down
- Fill Right — select edit, cells filled adjacent and to the right
- Fill Up — SHIFT + edit, cells filled adjacent and up
- Fill Left — SHIFT + edit, cells filled adjacent and to the left

Numbers can be entered into a cell by selecting the cell and typing the number. Numbers are automatically aligned to the right. Formatting commands allow the alignment and formatting of numbers. The Format drop-down menu contains eight format commands.

- Number — alters the number of digits and appearance
- Alignment — changes the position within a cell
- Font — changes the font and its size
- Border — adds borders and shading
- Cell — protects the contents of a worksheet
- Row Height — adjusts row height
- Column Width — adjusts column width
- Justify — displays text with the same width

Text can be entered into a cell as numbers are. The initial character of the text should not contain a mathematical operator and it cannot begin with an equal sign unless you use the following syntax:

="equal sign or mathematical operator"

After the text is entered, the commands under Format except for Number can be used to alter the appearance and alignment of the text within the cell.

The Cut, Copy, and Paste commands under Edit on the menu bar are important features for transferring the partial contents of a cell or the total contains a cell or cells to other parts of the worksheet. This is our next consideration.

Editing Worksheets

The Edit drop-down menu contains 12 edit commands. We have considered the Fill Right and Fill Down commands. We will now focus on Clear, Undo, Repeat, Copy, Paste, Cut, Insert, and Delete.

The Clear command is used when you want to clear the contents from a cell or a group of cells. When a cell is active, the delete key can also be used to clear the entire contents from that cell in one keystroke. When the Clear command is selected, a dialog box with four choices appears on the worksheet.

- All clears everything
- Formats clears only the formats
- Formulas clears formulas, numbers, and text
- Notes clears only the notes

The Undo command reverses the last command or cell entry. The Repeat command does just that.

The Cut and Copy commands place the contents of a cell or cells temporarily in memory for placement in another cell or group of cells. The Paste, Paste Special..., and Paste Link commands are only available immediately after the use of the Cut or Copy commands. The Copy command does just that. The Cut command not only copies the selected area but it deletes that area after the contents have been "pasted" in the new area. The Paste command places the formulas, values, formats, and notes in the new area. The Paste Special... command allows for the choices of pasting All, Formulas, Values, Formats, and Notes.

The Insert command allows you to insert a cell, a group of cells, a new row, or a new column. This command will produce a dialog box with the options of Shift Cells Right or Shift Cells Down. The Delete command will completely remove a cell, a group of cells, a row, or column from the worksheet. The Delete command will present a dialog box with the options of Shift Cells Left or Shift Cells Up.

The File Commands

After an initial worksheet has been opened, the commands on the drop-down menu under File on the menu bar are increased to 12. The New... and Open... commands are the same as previously described. The Close command closes the worksheet that is currently active. The Page Setup... command allows several options of which the most important are the choices of printing the Row & Column Headings and the Gridlines. Before printing for the first time, the Printer Setup... command should be selected in order to select the current printer and the printer settings. When you wish to print an active worksheet, you select Print... from these commands. Excel will present a dialog box with options on the number of copies and the pages to print.

Window Commands

The window commands allow you to open and close windows, to split windows into panes, and to arrange several windows into one window. As new worksheets are created or old ones opened, they are "stacked" on top of each other on the monitor screen. On the drop-down menu under Window on the menu bar, the titles of the worksheets are added to the bottom of this drop-down menu. To activate any particular window, you double click on that title. The Arrange All option allows you to arrange all of the open windows into one window.

Drop-Down Menus

An Alt followed by the underlined letter of the command category on the menu bar will expose the drop-down menu for that series of commands. The Esc key will retract a drop-down menu from the screen. A given command on the drop-down menu can be selected by keying the underlined letter for that command. You may also press the Alt key and use the left and right arrows to highlight the menu that you want to use. Press the down arrow key to pull down that menu. Select within the drop-down menu with the up and down arrows.

Other Important Commands

Two important commands occur under the category of Options. These commands are Display which allows you to show all of the formulas that are contained in the cells and Calculation which must be activated for use of the iteration technique. There are several other commands and operations that are considered under separate appendices. These are:

Appendix B Excel Error Values

Appendix C Graphing With Excel

Appendix D Excel Macros

Appendix E Saving Files With Excel

B

Excel Error Values

An error value is displayed in a cell when the formula in that cell cannot be calculated by Microsoft Excel. The error may or may not have occurred in that cell because it may reference a cell that produces an error value. You may have to trace the error values through a series of cells to locate the source of the original error. Error values always begin with the number sign (#). The error values are:

- #DIV/0!
- #N/A
- #NAME?
- #NULL!
- #NUM!
- #REF!
- #VALUE!

#DIV/0!

The #DIV/0! error value signifies an attempt to divide by zero. You may be using a reference to a blank cell or a cell that contains zero. You may have a formula that contains an explicit division by 0. In writing a macro, you may have used a macro function that returns #DIV/0! in certain situations.

#N/A

The #N/A error value means "No value is available." You may have omitted one or more arguments to a function macro. A formula that refers to a function macro that did not successfully function will return #N/A. In writing a macro, you may have used a macro function that returns #N/A! in certain situations.

#NAME?

The #NAME? error value is returned when Microsoft Excel doesn't recognize a name that you have used. Some of the common causes are: failing to define the name, misspelling the name, misspelling the function, omitting a colon in a range reference (=SUM(A1:A7)), and entering text without double quotation marks in the formula.

#NULL!

The #NULL! error value means that you specified an intersection of two areas that do not intersect. You may have used an incorrect range operator, or incorrect cell reference.

#NUM!

The #NUM! error value indicates a mistake involving a number. You may have used an improper argument in a function that requires a numeric argument. The #NUM! error value may also result from a computation that produces a number that is too large or too small to be represented by Microsoft Excel. When a function that uses an iteration technique can not find a solution, this error value is returned.

#REF!

The #REF! error value occurs when you refer to a cell that is not valid. You may have deleted or never entered information into cells that other formulas refer to. In writing a macro, you may have used a macro function that returns #REF! in certain situations.

#VALUE!

The #VALUE! error value occurs when you have used the wrong type of argument or operand. You may have entered text where a number or logical value is required. You may have referred to a range of cells for a function that requires a single value. You may have used an invalid matrix in one of the matrix worksheet functions. In writing a macro, you may have used a macro function that returns #VALUE! in certain situations.

C

Graphing With Excel

Microsoft® Excel can create charts from the worksheets that you develop. Excel has seven types of charts: area, bar, column, line, pie, scatter, and combination. After selecting the cells (refer to layout of the worksheet) that you want to include in the chart, select File and from the drop-down menu select New... to create a new chart. You then select Chart, which will create a new chart of your worksheet data. The default chart, column - option 1, will appear on the monitor screen. This chart is one of the forty four formats for the seven chart types that is available. The seven types of charts are listed on the drop-down menu under Gallery on the menu bar.

Layout of the Worksheet

Excel graphs data presented in many different forms. One design of a worksheet for graphing uses the first column for the independent variable, x, and the second column for the dependent variable, y. It is important to always leave the upper left corner blank of data or titles when producing a chart of this data. Worksheet C.1, illustrates the layout of the data. and Chart C.1, a line chart - option 1 graph of this data.

	A	B
1	X	
2		Y
3	1	5
4	2	7
5	3	9
6	4	11
7	5	13
8	6	15
9	7	17
10	8	19
11	9	21
12	10	23

Worksheet C.1

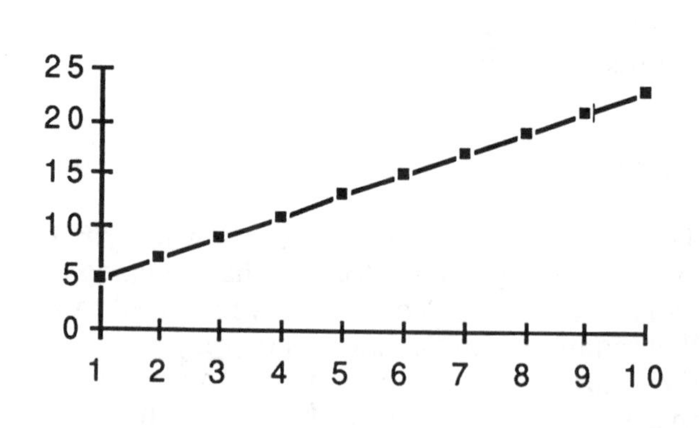

Chart C.1

Points and Line Charts

Select cells containing data to be plotted. Upper left corner is blank. Usually, three columns, independent variable, experimental dependent variable, predicted dependent variable.

Choose File, New, Chart, Gallery, Line..., 3, Chart, Add Overlay.

Adding Text

Horizontal axis
Chart, Attach Text, Category Axis, "text", "Enter", "click the mouse".

Vertical axis
Chart, Select Chart, "text", "Enter", place cursor on label, press the mouse button and move, "Enter", "click the mouse".

Chart title
Chart, Attach Text, Chart Title, "text", "Enter", "click the mouse".

Points and Scatter Charts

Select cells containing data to be plotted. Upper left corner is blank. Usually, three columns, independent variable, experimental dependent variable, predicted dependent variable.

Choose File, New, Chart, Gallery, Scatter..., 1, Chart, Add Overlay, Format, Overlay..., Scatter.

The Chart Menu

The most important commands on the drop-down menu under chart follow:

- Attach Text... enters chart title, axis or data labels
- Add Arrow adds an arrow to the chart, with or without the tip
- Add Legend adds chart legend
- Axes... specifies axes to display
- Gridlines... specifies gridlines to display
- Add Overlay divides chart into main and overlay

The Select Chart option activates the entire chart and allows the use of the Patterns... and Font... commands under Format.

The Format Menu

After you click the cursor on the horizontal axis, or the vertical axis, you may use the following commands:

- Patterns... changes patterns and colors of selected objects
- Font... changes font in selected cells
- Scale... changes scale of selected axis
- Main Chart... changes main chart settings

Activation by clicking on the chart title, value axis, or category axis on the actual worksheet, activates the following commands:

- Text... changes text settings of selected objects

A common procedure for all of these commands is the necessity of selecting the chart object before the command is selected. If this procedure is not adhered to, the commands on the drop-down menu under Format will not be available for use.

D

Excel Macros

A macro is a set of instructions that you can create for Microsoft Excel to follow. There are two kinds of macros, a command macro and a function macro. A command macro carries out a sequence of Excel instructions like you would normally select with keystrokes or the mouse. This book does not make use of this kind of macro. The book does use function macros, which compute a value. It is essentially a way to create mathematical functions that are not already built into Excel.

Function macros are used in Chapter 10 to define atomic orbital functions that are used for calculating molecular and hybrid orbital probabilities. These functions, if written completely out in a cell, would exceed the maximum number of characters allowed in a formula. The function macros are set up in the function macro sheet, O.XLM, and these are used in the formula to build up the complete expression staying within Excel's limit of the number of characters in a cell.

Chapter 10 has a complete description of setting up, naming, and using the function macro worksheet used in that chapter. As a reminder, always open or create the function macro sheet, O.XLM, first, followed by opening or creating the orbital plotting worksheets.

E

Saving Files With Excel

You may wish to save your worksheets and charts onto a disk for later use. This may be to finish a worksheet that you did not have time to complete or to use the worksheet for additional data. You must first decide which disk you want to use to save your files, the hard disk or a floppy disk. You may need to consult your instructor as to which would be best or allowed for your situation. The hard drive is the simplest to use but you may not be able to use it if a number of other students are saving files also. In that situation, a floppy disk that you can take with you is the safest.

Hold down your mouse button with the arrow on File on the left part of the menu bar at the top of the screen. While continuing to hold down the mouse button, drag the arrow to Save As... and release the button. A dialog box will appear. Type the name under which you wish the worksheet or chart to be saved in this box and click on OK (Save on the Macintosh). This will save the file on the hard disk drive if you haven't changed that as the default drive since starting Excel. To save your work to a floppy disk on an IBM type machine, insert a formatted disk into drive A and follow the same procedure as for the hard drive, except when you name the file, start the name with *a:* . For example, to save a worksheet named *gas*, use the name *a:gas*. If you are using a Macintosh, click on Drive until you get the disk name on which you wish to save the file before clicking on Save.

If you are saving a chart associated with a worksheet, save the worksheet first so when you save the chart it will know what you called the worksheet and be linked to the worksheet.

After you have saved a file once, you can use Save under File on the menu bar. This will use the same name and disk drive as you previously used.

To retrieve files that have been saved on disk, select Open under File on the menu bar. Click on the directory or disk drive under which your file is stored (click on Drive on the Macintosh). Find the file name you wish to open under Files, using the scroll bar if necessary, and click on that file name. Then click on OK (Open on the Macintosh) and the file should be loaded. Load any related charts in the same manner. You can then adjust the window(s) as desired or select Window, Arrange All.

Open a worksheet first if a chart is linked to it. Then open the chart. Use Window, Arrange All to display both at the same time. For Chapter 10, Molecular and Hybrid Orbitals, always open the function macro sheet, O.XLM, first, followed by the orbital plotting worksheets.

F

Successive Approximations

Modern spreadsheets have the capacity for solving circular logic or iterative procedures. Microsoft's® Excel has the capability of processing algorithms that have formulas that depend on each other. With an iterative procedure it is possible to find to any desired degree of accuracy, the real roots of many polynomial equations. All that is required is the mathematical logic and a few steps to set Excel for this type of computation. After the formulas are entered in the cells of the spreadsheet, Excel will respond with the message "Can's resolve circular references." Select Options from the menu bar and choose Calculation. Check the box for Iteration. If the calculations are not produced at this point, the following procedure must be followed.

Use the cursor to select the cell that contains the value that the circular logical is seeking. This is cell B7 in the example given below. Then click the cursor on the formula bar (use care so that you do not disturb the formula that is shown), and then select and click on the square that contains the check. This will initiate a new series of iterations in search of a solution that converges.

	A	B	C
1	Hydrofluoric Acid		
2			
3	0.00068	= Ka	
4	0.001	= [HF] init	
5			
6		[H+]	pH
7	iteration method	0.000552	3.26
8	approximate	0.000825	3.08

	B	C
1		
2		
3	="= Ka"	
4	="= [HF] init"	
5		
6	[H+]	pH
7	=(K*(A4-B7))^0.5	=-LOG(B7)
8	=(K*A4)^0.5	=-LOG(B8)

Worksheet F.1 and Worksheet F.1F

The spreadsheet shown above computes $[H^+]$ for a weak acid. This calculation involves solving a quadratic equation. There are two solutions for this quadratic equation and only one of these is of significance to the chemist. If the spreadsheet converges on the negative root, it is possible to alter the spreadsheet or change the initial concentration so that the positive solution is computed. The root of chemical significance must be positive.

A spreadsheet that utilizes a technique for inserting in the iteration cycle a new or different guess for the root is illustrated by Worksheets F.2 and F.2F. This example involves finding the roots of a second degree equation, $x^2 - x - 2 = 0$. One of the roots is a positive two and the other is a minus one. The conditional or logical statement in cell B8 inserts you new guess when you enter a positive number in cell B5 as the initial guess and the text, "new," in cell C5. This computation cycles 100 times with the end result being the insertion of you new guess in cell B8. You then enter the text "start" in cell C5 and the iterations proceed to converge on the positive root, two. If you use the same process and enter a negative number for the initial guess, the iterations will converge on the negative root, minus one.

	A	B	C	D
1	Newton - Raphson Iteration Technique			
2				
3	find roots ->	x^2-x-2=0		1. guess
4				2. "new"
5	initial guess ->	11	start	3. "start"
6	f(x)	-0.0001022		
7	f'(x)	1.99993188		
8	x	2.00001703		

	B
1	Raphson Iteration Technique
2	
3	x^2-x-2=0
4	
5	11
6	=B8^2-B8-2
7	=2*B8-2
8	=IF(C5="new",B5,B8-B6/B7)

Worksheet F.2 and F.2F Initial Guess Method

For the iteration method, you must select Calculation... from the drop-down menu under Options. Then click on the Iteration box and accept or alter the default values of 100 iterations and a maximum change of 0.001. You may accept automatic calculations or choose to use the manual calculation mode. A calculation is initiated by the keystrokes, Alt followed by Shift + O followed by keying the letter, M. An Alt followed by the underlined letter of the command category on the menu bar will expose the drop-down menu for that series of commands. The Esc key will retract a drop-down menu from the screen. A given command on the drop-down menu can be selected by keying the underlined letter for that command.

G

Newton-Raphson Method

A very useful method for finding the roots of nonlinear polynomial functions is the Newton-Raphson iterative procedure. In general, a nonlinear polynomial function is a curve on a graph when the values of the function are plotted on the vertical axis and the values of the independent variable, x, are plotted on the horizontal axis. The object of the procedure is the location of the point, the value of x, at which the curve crosses the horizontal axis. The curve may cross the horizontal axis at more than one point. The expectation is that there will be n roots or solutions for an nth degree polynomial function.

$$f(x) = a_0 \cdot x^n + a_1 \cdot x^{n-1} + \cdots + a_{n-1} \cdot x + a_n$$

Not all of the roots of a polynomial function are always real (the term, a real root, implies that the function crosses the x-axis). Roots of polynomial functions are imaginary in some cases and these are always present as pairs or conjugate roots, a + bi and a - bi. Imaginary roots are of no interest for chemical systems. Even some real roots are meaningless in terms of the chemical system, a negative molar concentration as an example. Solutions for polynomial functions that describe chemical systems are made much easier because many solutions do not make sense in terms of the chemistry. Because of the boundaries on the mathematical solutions imposed by the chemistry, the design of spreadsheets for finding these solutions is much easier than the general problem of finding all of the roots for a given polynomial function.

The Newton-Raphson iterative procedure starts with an initial value of x_0 that may be no more than a guess. The value of the function and the slope of the function at that specific value of x is calculated. The method uses the value of the function (point on the curve) and the slope of the tangent to that point to develop an equation for the slope. The equation for the slope is then evaluated for the point at which it crosses the horizontal axis, which produces a new value of x, x_1. Each successive point is calculated as the point where the tangent at the previous point intersects the x-axis. With some luck the iterative procedure converges on a desired root. The Newton-Raphson equation for finding successive values for x is

$$x_1 = x_0 - \frac{f(x_0)}{f'(x_0)}$$

The term $f'(x_0)$ represents the derivative of the function which is a calculus technique for finding the tangent to the curve at the point where $x = x_0$. Polynomial functions are the only equations needed for solving chemical systems of interest to us.

The general form of the derivative of a polynomial term is

$$f(x) = C \cdot x^n$$

$$f'(x) = n \cdot C \cdot x^{n-1}$$

An example of this type of process will illustrate the ease with which a derivative of polynomial is produced.

$$f(x) = 3 \cdot x^4 + 2 \cdot x^3 + 5 \cdot x^2 + 6 \cdot x + 7$$

$$f'(x) = 12 \cdot x^3 + 6 \cdot x^2 + 10 \cdot x + 6$$

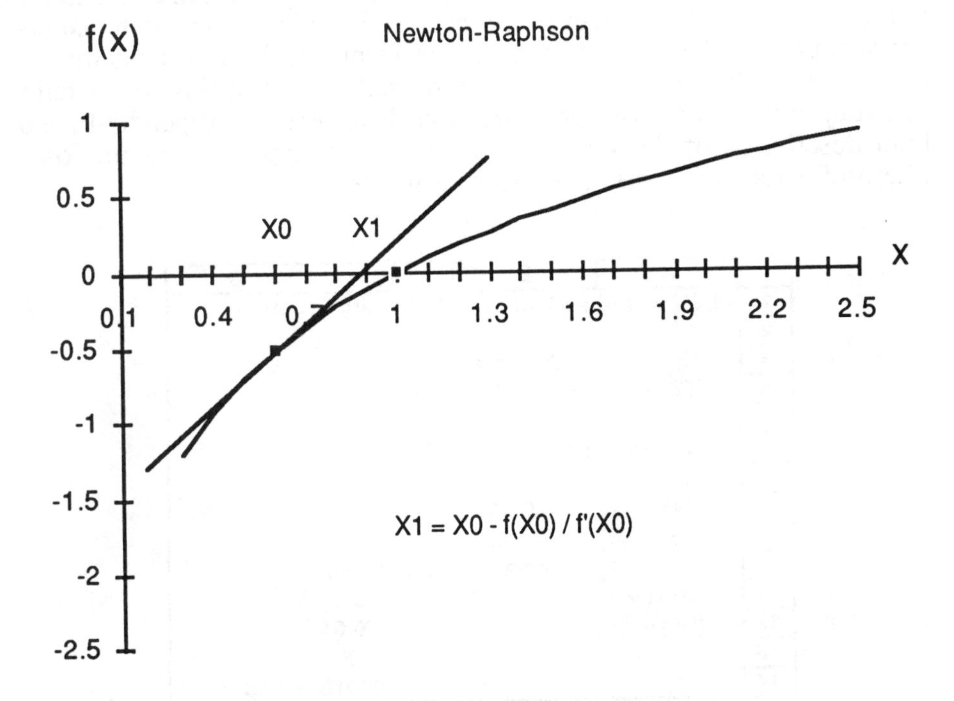

Chart G.0 Finding Roots of a Function

Solubility of $PbCl_2(s)$ in Solutions Containing Chloride

The solubility of $PbCl_2(s)$ in aqueous solutions that contain added chloride serve as an example of the Newton-Raphson technique. The solubility product for this equilibrium is large enough for the dissolution of $PbCl_2(s)$ to contribute significantly to the equilibrium concentration of chloride. The chemical reaction and equilibrium expression are

$$PbCl_2(s) = Pb^{2+}(aq) + 2\,Cl^-(aq)$$

$$1.6 \times 10^{-6} = K_{sp} = [Pb^{2+}]\cdot[Cl^-]$$

$$K_{sp} = x\cdot([Cl^-]_0 + 2\cdot x)^2$$

$$\text{where } [Pb^{2+}]_{eq} = x \text{ and } [Cl^-]_{eq} = [Cl^-]_0 + 2\cdot x$$

Worksheet G.1 illustrates the equilibrium concentrations for both lead(II) and Chloride. With a solution that is initially at a concentration of 0.0200 M in chloride, the dissolution of the solid, $PbCl_2$, produces an equivalent amount of chloride ion. The resulting equilibrium concentration of lead(II) is much lower than would be calculated by an approach that does not account for the chloride ion that is furnished by the dissolution of $PbCl_2$. Worksheet G.1F presents the formulas used in this computation. The subject of *Successive Approximations* is covered in Appendix F, and a brief description of the steps necessary for this type of computation by Microsoft® Excel follows the two spreadsheets.

	A	B	C	D
1	PbCl2 = Pb2+ + 2 Cl-		Solubility Product	
2				
3	Ksp =	0.000016		
4	[Cl-] =	0.02		
5				
6	Newton -	Raphson Iteration		
7	f(x)	1.2826E-13		
8	f'(x)	0.0032		
9	x	0.01		
10		Initial	Equilibrium	
11	[Pb2+] =	0	0.01	
12	[Cl-] =	0.02	0.04	
13				
14			0.000016	= Ksp

Worksheet G.1 Solubility of $PbCl_2$ in Chloride Solutions

	B	C	D
7	=4*x^3+4*Cl*x^2+Cl^2*x-K		
8	=12*x^2+8*Cl*x+Cl^2		
9	=x-Y/DY		
10	Initial	Equilibrium	
11	0	=x	
12	=Cl	=Cl+2*x	
13			
14		=C11*C12^2	="= Ksp"

Worksheet G.1F Formulas for Worksheet G.1

Entering formulas that have specific cells defined by an alphanumeric representation can be a bit tricky. If you enter a formula with a defined cell as a variable and that cell has not been defined, the error value "#NAME?" appears. Continue to enter formulas in all of the cells and then choose Formula from the menu bar. Select Define Name and define each of the cells as needed. If the error value "#NAME?" remains after you have entered all of the formulas and definitions, proceed as follows. For every cell that contains that error value, select one and click the cursor on the formula bar, followed by selecting the check mark. Repeat this until you have processed all of the cells that contain the error value. You are then ready to set the spreadsheet for an iterative procedure.

With circular logic, Excel returns the message, "Can't resolve circular references." Choose Options from the menu bar and select Calculation. Check Iterations and set the Maximum Change to 0.0001. You may have to repeat the process used in the preceding paragraph. Select the cell containing x, and click on the formula bar followed by selecting the check. Repeat this process on the other cells involved in the circular logic. The spreadsheet should function at this point. If the error values remain after this process, one or more of the formulas contains an incorrect entry.

Formula List G.1 lists cells that are defined by an alphanumeric characters. This spreadsheet uses zero as the first approximation for the root.

Formula List G.1 Formulas For Worksheet G.1

Cell B3	Defined as K
Cell B4	Defined as Cl
Cell B7	Defined as Y
Cell B8	Defined as DY
Cell B9	Defined as x

Index

Q

R

S

T

V

W